my **revision** notes

WJEC B GCSE
GEOGRAPHY

The publishers would like to thank the following for permission to reproduce copyright material:

Acknowledgements

P.19 Figure 3, from World DataBank (2013), reproduced by permission of The World Bank Publications; **p.21** 'Homes for Wales - Housing Bill', from www.Wales.gov.uk (7 December, 2012); **p.33** Figure 11, from State of the World's Cities, 2010-2011: Bridging the Urban Divide, UN – Habitat; **p.39** Press Release, adapted from http://www.plunkett.co.uk/newsandmedia/news-item.cfm/newsid/655; **p.43** Steven Morris, 'Second homes at Rock, Cornwall', adapted from The Guardian (23 October, 2012), copyright Guardian News & Media Ltd 2012, reproduced by permission of Guardian News & Media; **p.86** 'Sustainable development' from United Nations, Brundtland Report (1987); **pp.96** 'Oxfam calls for a rapid scale up of aid effort from governments and individual donations to tackle Sierra Leone's escalating cholera outbreak', Oxfam News Bulletin (23 August, 2012); **p.101** 'Case Study – Water, sanitation and hygiene education in Ghana', Oxfam/WaterAid, http.www.oxfam.org/en/development/ghana/hygiene-education, © Oxfam International.

Maps on **p.66, and p 77** reproduced from Ordnance Survey mapping with the permission of the Controller of HMSO, © Crown Copyright. All rights reserved. Licence number 100036470.

Photo credits

P.19 © Still Pictures/Robert Harding; **p. 24** *all* © Stuart Currie; **p.32** © IOM/MPW Photography Project 2007 – Lerato Maduna; **p.36** © Stuart Currie; **p.37** *all* © Stuart Currie; **p.39** © Plunket Foundation; **p.41** © Stuart Currie; **p.47** © Stuart Currie; **p.54** © 2006 TopFoto/Jon Mitchell; **p.56** © NASA/Goddard Space Flight Centre; **p.64** © Michael S. Yamashita/CORBIS; **p.66** *l* © Eric Foxley, *r* © Stuart Currie; **p.67** © Stuart Currie; **p.68** *all* © Stuart Currie **p.69** © Stuart Currie; **p.70** © Colin Lancaster; **p.71** © Stuart Currie; **p.72** *all* © Stuart Currie; **p.82** *t* © Getty Images, *b* © AFP/Getty Images; **p.96** © Tugela Ridley/epa/Corbis; **p.98** *t* © 2012 Getty Images, *b* © 2013 Getty Images; **p.101** *tl* © Oxfam, *tr* © WaterAid, *b* © John Spaull/WaterAid; **p.107 and p.114** © LARRY MACDOUGAL/AP/Press Association Images.

Orders: please contact Bookpoint Ltd, 130 Milton Park, Abingdon, Oxon OX14 4SB. Telephone: +44 (0)1235 827720. Fax: +44 (0)1235 400454. Lines are open 9.00a.m.–5.00p.m., Monday to Saturday, with a 24-hour message answering service. Visit our website at www.hoddereducation.co.uk.

First published in 2013 by
Hodder Education,
an Hachette UK company
Carmelite House, 50 Victoria Embankment,
EC4Y 0DZ,London

Impression number 10 9 8 7 6

Year 2017 2016

Typeset in Datapage (India) Pvt. Ltd.

Artwork by Datapage (India) Pvt. Ltd.

Printed and bound in Spain

A catalogue record for this title is available from the British Library

ISBN 978 1 444 193909

my **revision** notes

WJEC B GCSE GEOGRAPHY

Stuart Currie

Contents and revision planner

Theme 3 Uneven Development and Sustainable Environments

Problem solving

Get the most from this book

Why should I use this book?

These revision notes will aid your revision for the WJEC B GCSE Geography specification to help you get the best possible result in your exams.

There is a great deal you need to know in order to obtain a result you will be proud of in this subject. You not only need to know the subject of geography well but, of equal importance, you will also need to know how to use all of your geographical abilities to get the best out of the examination experience. This involves understanding exactly what the examiner wants of you and being able to provide this in the examination situation.

But don't panic – that's where these notes come in! They take you back through all the main areas of content for your course and will also train you in how to use this information to get the best out of your exams. By the time you have finished you will know almost as much as your examiners – not a bad position to be in!

So, please don't ignore the opening pages of these notes. They are the key to getting the best out of the rest of the book and, as a result, obtaining the best possible result for *you* in geography. There are many candidates who are very good geographers but never quite develop the ability to show this in the examination room. Read on and get involved in the activities to ensure that you are not one of these people.

Using these revision notes

These revision notes are not all you need to gain examination success. They have certainly not been written with the intention of replacing your teacher, the most important resource you have.

Most of you will have been studying geography since, at least, entering secondary school and will have learned a great deal in that time. You will probably also have notes in exercise books and files that will help you prepare for your examinations. These notes and your teachers are the *real* key to your success in the examinations.

These revision notes will, therefore, help you to make sense of your own notes and train you in the art of how to use your own geographical competencies to respond to the variety of tasks your examiners will set before you. Your teachers will also be working hard with you to ensure your examination success and it is my intention that these revision notes help in this process.

Unlike many revision guides, this book does not contain huge amounts of facts. These you already have. It does, however, help train you in those abilities you will require to get the best out of the examination experience; how to respond in the examination to the demands for you to apply your geographical knowledge, understanding and skills to new situations.

Features to help you succeed

These revision notes will help you to revise for the WJEC B GCSE Geography specification. It is essential to review your work, learn it and test your understanding. Tick each box when you have:

- revised and understood a topic
- checked your understanding and practised the exam questions.

You can also keep track of your revision by ticking off each topic heading in the book. You may find it helpful to add your own notes as you work through each topic.

You can download a revision planner from **www.therevisionbutton.co.uk/myrevisionnotes** to plan your revision, topic by topic.

Tick to track your progress

Exam tip

Throughout the book there are exam tips that explain how you can boost your final grade.

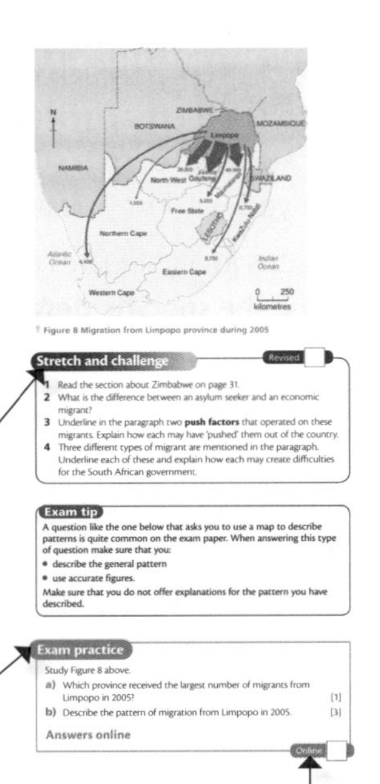

Key term

Essential key terms are defined on the page. Go online to see the full glossary at **www.therevisionbutton.co.uk/myrevisionnotes.**

Knowing the basics

These activities are designed to help you understand and revise key principles and concepts you need to know for the exam. Check your answers in the back of the book.

Stretch and challenge

These activities are designed to help you develop the reasoning skills expected of an A grade/high performing student.

Exam practice

Sample exam questions are provided for each topic. Use them to consolidate your revision and practise your exam skills.

Go online

Go online to check your answers to the exam questions at **www.therevisionbutton.co.uk/myrevisionnotes.**

Getting to know the specification

The following is an overview of the three themes that make up the GCSE Geography for WJEC B specification. The first key to doing well in your examinations is to develop an understanding of each of these.

- Each large area of content is called a 'theme'.
- Each theme is tested through the two 'units' or examinations.
- Unit 1 tests Themes 1 and 2.
- Unit 2 tests Theme 3.
- Unit 2 also tests your ability to solve a geographical problem.

Theme 1: Challenges of living in a built environment	Theme 2: Physical processes and relationships between people and environments	Theme 3: Uneven development and sustainable environments
• Variations in quality of life and access to housing • Access to services and changing service provision • Urbanisation • Planning issues in built environments • Rural change and planning issues	• Weather and climate • Ecosystems • The issue of desertification • River processes and landforms • Coastal processes and coastal management	• Employment structures and opportunities • The location of economic activities • Economic activity and the environment • Development • Interdependence • Development issues and water

A case study overview

You need to have studied a total of 18 case studies during your geography course. There are six of these for each of the three themes, and your examiners will test any two of each set of six during any examination session. If you don't know your case studies and how to apply them to the questions asked you will lose valuable marks.

One of the main reasons for candidates not getting the grade they expect is not bothering to revise the **specific detail** of the case studies they have been taught.

The tables on pages 9–11 show the case studies you will need to know for each theme. Although the ideas for each case study are the same for all candidates, the geographical locations you have studied them in have been chosen by your teacher.

Complete the middle column of each table and add brief notes about the case studies *you* need to learn. Then complete each final column with information about where you can find all your notes on each case study. To do this you will have to make sure that your work is well organised. As you can probably imagine, this is something that is best done as you go through your course. Don't leave everything until the last minute! Organising your work as you go will keep you on top of things and help you make up any gaps as they happen.

The tables are just a brief reminder of your case studies. Ideas for remembering the specific detail will be discussed later.

Are you naturally untidy? If so, train yourself in the art of keeping tidier notes. If your work is in exercise books, number these and write a brief contents list on the inside front cover of each book. Create something similar for loose-leaf notes and don't forget to number the pages.

Case studies: Challenges of Living in a Built Environment

Case study	Brief description of my case study	Where are my notes?
1 Housing in an urban area • Under what terms is housing occupied? • What determines people's access to housing – opportunities – constraints? • What patterns does this produce?	Location: Coventry Description: St. Michaels Inner City Wainbody Outer Suburbs	
2 Retail service provision • How are the services distributed in an urban area? • How do the services vary across the urban area? • How is urban retail service provision changing?	Location: Description:	
3 Other service provision • How are the services distributed in an urban area? • How do the services vary across the urban area? • What affects access to the service for different groups of people?	Location: Cinemas Description: Odeon CBD Warwick Arts Centre Showcause	
4 Rural to urban migration • What is the pattern of migration? • Why has the migration taken place? • What are the impacts on the area the migrants come from? • What are the impacts on the area the migrants go to?	Locations: Description: To Kaba Kenya → Nairobi	
5 A planning issue in a residential area • What changes are planned? • Why has it caused a conflict? • What are the views of different groups of people on the issue?	Locations: Sites + Services Description: in Nairobi Kenya or Thomas King House, Coventry	
6 Leisure use of a rural area that causes conflict • What leisure use(s) cause the conflict? • What is the nature of the conflict? • How is the conflict being managed to balance change and sustainability?	Location: Description: Peak District – dark Peak used for walking	

Case studies: Physical Processes and Relationships between People and Environments

Heat wave 2003- UK and Europe

Case study	Brief description of my case study	Where are my notes?
7 An extreme weather event • What were its causes? • What were its effects on different groups of people? • How did people respond to this hazard?	Location: Description: Heat wave	
8 An ecosystem • What natural processes take place in the ecosystem? • How does the ecosystem benefit people? • What is the impact of human activity? • How is human activity being managed?	Location: Description: savannah	
9 Desertification • What are the natural causes of desertification? • What are the human causes of desertification? • What are the effects of desertification? • How might desertification be managed?	Location: Description: sahel	
10 A river flood • What are the natural causes of the flooding? • What are the human causes of the flooding? • What are the effects of flooding – on people – on the environment?	River system: Description: Boscastle, Bangladesh	
11 Management of river flooding • What attempts are being made to prevent the flooding? • What attempts are being made to protect from the effects of flooding? • How successful are these attempts? • What are the views of different groups of people on the issue?	Location: Description: Boscastle	
12 Coastal processes, landforms and their management • What processes affect this coastline? • How do they create landforms? • How are the processes being managed? • How effective is this management?	Coastline: Description: Hurst Castle	

Case studies: Uneven Development and Sustainable Environments

Case study	Brief description of my case study	Where are my notes?
13 Changing location of a secondary or tertiary industry • How did it change location? • Why did the change occur? • What were the social and economic impacts of the change?	Secondary / tertiary (circle) Location: croft quarry Description:	
14 A multinational company • Why has it located where it has? • What are the economic effects on these places? • What are the social effects on these places?	Location(s): pyson Description:	
15 Impact of an economic activity on the environment • What are the causes of the environmental damage? • What are its effects on the environment? • What strategies are being used to manage the environment?	Location: Description: dark peak	
16 Managing the effects of climate change • What strategies are being used to manage the effects of climate change at the – local – national – international scales? • What do different groups of people think about these strategies?	Local: National: International:	
17 A trans-boundary water issue • Which countries or regions are involved? • What is the issue? • What attempts are being made to resolve the issue?	Countries or regions: Description:	
18 International aid • Why is the aid needed? • What is the nature of the aid? • How effective is the aid?	Countries involved: Description: smokeless cookers	

Assessment through WJEC specification B

The route to the examination

This will differ depending on the school or college you attend.

Most courses start at the beginning of Year 10, although it is possible that you may have started earlier, at some time in Year 9, or later, possibly even as late as the beginning of Year 11. Whatever your starting point, the examination assumes that you are a 16-year-old at the time.

Before the final exam, you will have completed one item of school-based work known as the 'Enquiry'. This is a controlled assessment item. It will be set and marked by your teachers who will then send a sample of their marking to a moderator who makes sure they have applied their marks according to the agreed national standard. You will know the 'unmoderated' mark for the Enquiry before you enter the examination room.

It is very important that your mark for the Enquiry is one that shows your full geographical abilities. It is, therefore, important that you work as hard for that as for the preparation for the written papers.

You will sit the written papers in two sessions known as Units. They are usually a few days apart.

Unit 1 will test both Theme 1 and 2. Most of the questions you will answer will be based on resources like maps, graphs, diagrams and photographs that are a part of the paper. However, you will also need to know case studies 1–12 for this Unit.

Unit 2 will test Theme 3. The structure of the paper is similar to Unit 1 and for this paper you will need to know case studies 13–18. It will also test your ability to solve a geographical problem. So, after 30 minutes, your Theme 3 paper will be taken from you and you will be given the second paper in this Unit, the 'problem solving' paper. There is much more guidance for this towards the end of this book.

The structure of the entire examination (the big picture) is shown in the table below.

The big picture

Name	Nature of assessment	Time	Proportion of total mark
Controlled assessment	The Enquiry	up to 18 hours + 1 day field study	25%
Unit 1: Foundation or Higher Tier	Theme 1	1 hour	15%
	Theme 2		15%
Unit 2: Foundation or Higher Tier	Theme 3	30 minutes + 1 hour 30 minutes	15%
	Problem solving		30%

For geography, you will be entered for either the Foundation or Higher Tier written paper. In many cases the choice is an obvious one but the information below may help if you and your teachers are in doubt.

Foundation or Higher Tier? The facts

- You can get a Grade C by either route.
- The Higher Tier exams make greater demands on your language skills.
- The Foundation Tier exams organise your answers more for you.
- Entry by the Foundation Tier route is not a sign of failure.
- Higher Tier entry is essential for future A Level studies in Geography.

Making your mind up ...

- Consider your level of entry early in your course.
- How might this triangle help you decide?

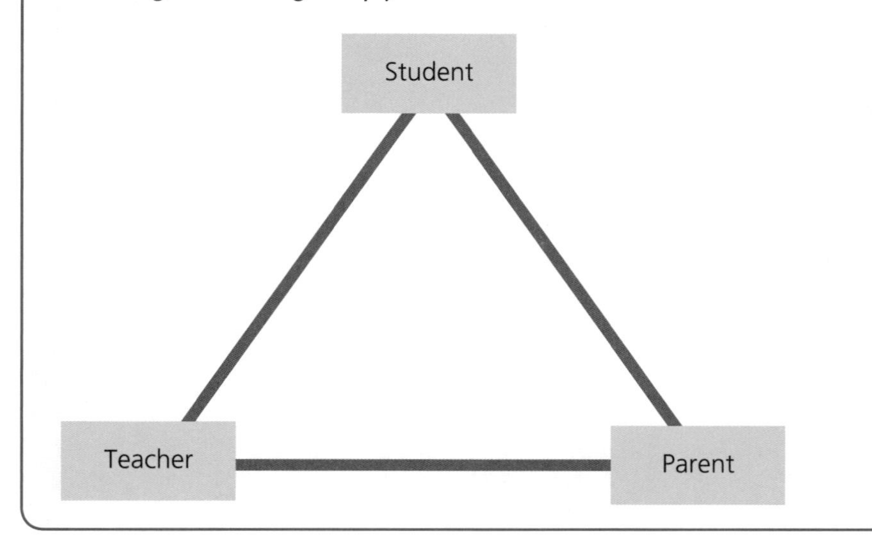

Making the exams work for you

It's worth shouting out loud, *'your examiners really want you to do well'*. They use quite a few different ways to make sure that you provide them with the answers that they want from you. This can only happen, though, if you play their game.

They expect you to:

1. Read all the information provided on the examination paper. There may be no questions to answer on its front page, for example, but there will be instructions and other information to help you.
2. Manage your time carefully. You must complete the paper if you are to do well. Your examiners help you by telling you how many marks are available for each question and by giving you the number of lines they feel you will need for your answer.
3. Do everything that is asked of you. Your examiners use 'command words' to tell you what sort of answer they want from you. These words always mean the same thing so you must always respond to them in the same way. Practise this throughout your course.

Command terms used in examinations

The table below gives the main command terms that will be used in your Geography exam. It also gives their meanings. Your task is to match each term to its meaning. One has been done for you. When you have finished, check your answers with the list on page 17.

Command word	Meaning
1 Circle	**A** Give an *accurate* figure by reading a graph or map. Do not just give an estimate.
2 Complete	**B** Create a sketch map or diagram to help explain a feature, describe a location or show your plans for an area.
3 List	**C** Place a circle round the correct answer from a list of alternatives.
4 Name	**D** Say what is similar and/or different between two pieces of information.
5 Locate	**E** Similar to 'explain' but expects you to provide the information from your own knowledge rather than using that provided by the exam paper.
6 Measure	**F** State where a place is.
7 Draw	**G** Explain why you have made a decision.
8 Describe	**H** Write down more than one feature you are asked to give from looking at a map, photograph or other resource.
9 Explain/Give reasons for …	**I** Fill in gaps in something like a graph or a sentence.
10 Suggest	**J** Just write what a feature is, e.g. the 'motorway M6' on an O.S. map.
11 Compare	**K** State why something you have described exists or has happened.
12 What is meant by …?	**L** Just say what you see. It may be a scene in a photograph or pattern on a map or graph.
13 Justify	**M** Give a definition of a geographical term.

Describe

Just write what you see. You will normally be asked to describe a photograph, map or graph. Look at the mark allocation to work out how much detail you need to give. Remember, do not *explain* anything for these questions.

What is meant by … ?

You are being asked to define a geographical term. You need to learn the key terms and definitions. Use the definitions in these revision notes to help you. Don't give an example instead of a definition; the examiners want to know that you have understood what the term means.

Explain/Give reasons for …

These questions test your knowledge and understanding. You are being asked to say *why* something you have already described is happening. Use 'because' to help you answer these questions.

There are often 2 marks awarded for giving just one reason. For these questions, you will be expected to give a simple statement and an elaboration. To write the elaboration, ask yourself 'So what?'

Suggest

This is similar to 'Explain' but tells you that you are expected to bring in ideas and understanding of your own, and therefore the ideas required are not provided in the exam paper.

Compare

When you see a question asking you to compare, you should write what is similar and what is different between two pieces of information. You should make use of words like 'whereas' and 'in contrast' to help you make comparisons.

Measure

You may be asked to measure something on a map or graph. Don't guess – measure accurately using the scale provided.

Effective revision

The only person who can decide how to revise most effectively is *you*. There is a huge variety of techniques and some will suit some people more than others. The following questions may help you to decide your most effective method of revision.

	Yes	No
Do you need complete silence to revise?		✓
Does music help to cut out outside noises?		✓
Can you concentrate for long periods of time?	✓	
Is your attention span short?		✓

Now think about your answers to the following questions:

- For how many different exams do you have to revise?
- Where is Geography in the exam timetable?
- What parts of my social life are essential?
- What can I give up to make time for revision?

There is just one rule when answering these questions – be totally honest with yourself. There is a very long period of time between your examinations and the results day. You will only enjoy this time fully if you have completely prepared yourself for the examinations and you can honestly say that you could not have done better.

So, now you are armed with this vital information, create your own customised revision programme that:

- starts early enough
- balances work and pleasure
- suits *your* concentration span
- is realistic in the demands it places on *you*
- takes place in conditions that suit *you*
- builds in rewards.

Finally, remember just three more points:

- Your teacher is there to help and will welcome questions.
- There may be Geography revision lessons offered during your study leave time – attend them!
- Everyone realises the pressure you are under. If you feel, at any time, you are not coping be sure to talk to someone about it.

There is a downloadable revision timetable on **www.therevisionbutton.co.uk/myrevisionnotes** which will give you a plan for a 12-week revision period. Complete this to give a focus to your revision activities. Use marker pens or simple ticks to track your progress.

Active revision

Your revision can be either *active* or *passive*. Passive revision involves just reading your notes and is something that is only likely to work over very short periods of time. After this the mind begins to wander and all sorts of outside influences will get in the way of effective revision, for example staring at posters on your wall or listening to noise coming from outside your room.

On the other hand, active revision involves you actually *doing* something. This action is likely to help you to maintain your concentration at a reasonably high level, and can often result in you also producing something that will be helpful later in the revision process.

Active revision techniques

Case study revision cards
Make revision cards for the main case studies you hope to use in the examination. Divide each card into areas for the separate features of the casestudy shown in the left columns of the tables on pages 9, 10 and 11 of these revision notes.

Remembering maps and diagrams
Well-drawn sketch maps and diagrams are always welcomed by your examiners. Try to memorise some of these and then attempt to redraw them. Compare your redrawn maps with the originals.

Drawing labelled diagrams
Link features affecting or influencing a particular feature by drawing spidegrams. This is useful for such ideas as influences on quality of life, for example local service provision or the effects on an area of river flooding.

Key term revision cards
Create cards to test yourself and your friends on some of the key terms needed for success in a Geography examination. Produce one set of cards of the terms and another set with their definitions. Use them as a simple matching exercise or a game of Geography 'snap'. You might consider creating a set for each of the three themes.

Each of these strategies for revision success is looked at in greater detail as you work through this book.

And finally ...

The big day has come, you have revised well and there is nothing that can get in the way of your success. Or is there?

Ensuring you have this information will further help you to succeed by keeping you in control of the situation. Complete this checklist for your Geography exam.

You should:

<table>
<tr><td rowspan="2">• reach the exam room in plenty of time</td><td>Unit 1
date:
time:</td></tr>
<tr><td>Unit 2
date:
time:</td></tr>
<tr><td>• know your centre and candidate number</td><td>Centre:
Candidate:</td></tr>
</table>

You should also:

- listen carefully to your invigilators
- carry spare writing equipment
- read the front page instructions carefully
- use your time well
- answer all questions – don't leave gaps.

Discuss all these points with friends, teachers and parents.

How will each point help you stay in control?

Are there any other ideas that will help you?

Questions: just what does the examiner want from me?

- Your examiners have designed the questions carefully to encourage you to show your geographical abilities as fully as possible.
- A few words the examiners feel are very important may be in bold or italics.
- You may wish to highlight other words if it helps you to understand the question.
- Look at the question below. A candidate has underlined the words she thought were important to her full understanding of the question. The messages she got from each area of highlighted text are shown in the speech bubbles.

Description is not enough. I must give reasons for the damage.

A really detailed explanation of one reason isn't good enough. Neither can I give a large list of different reasons.

Explain two ways in which people may damage rural areas they visit.

A tight focus. I must write only about how the damage is caused and not stray into attempts to prevent or manage it.

I can't write about just *any* area. It must be an area of countryside.

You may not wholly agree with the underlining by the candidate above. You may also feel that highlighting using a marker pen is more useful to you.

1. Highlight the pieces of text you consider most important in the three questions below.
2. Write a brief description of the messages each piece of highlighted text gives you about how to answer the question.

'Suggest advantages and disadvantages of building on greenfield sites.'

'Explain how **fair trade** might help poor countries with their development.'

'Compare the purposes of short-term emergency aid and long-term emergency aid.'

Answers for Exam tip: Command words (page 14)

1C; 2I; 3H; 4J; 5F; 6A; 7B; 8L; 9K; 10E; 11D; 12M; 13G

How do quality of life and standard of living differ?

How can quality of life be measured?

Revised

Many factors affect people's **quality of life**. Some, but not all, of these can be measured. For example, the amount of money you own may be measured but the influence of your friends on your life cannot.

The happiness, well-being and satisfaction of a person. | **Quality of life** ? Standard of living | Influences on the lives of people that can be measured.

The area in which you live will have both positive and negative effects on *your own* quality of life. Your immediate **neighbourhood**, your **school** and **access to shops, transport, leisure facilities** and **places of worship** may be major influences on it.

Stretch and challenge

Choose one service in your local area that has positive effects on one group of people and negative effects on another group. Explain the contrasting effects of the service on these people.

Revised

Knowing the basics

Revised

1 Complete a copy of the spidergram in Figure 1 to show the influence of where you live on your quality of life.

a) School and transport are listed on the spidergram. Show how these affect you in both positive and negative ways. Use different colours.

b) Add the other influences shown above and develop them outwards to show both positive and negative impacts on *you*.

c) Add cross links between the main influences to show how some influences work together to affect your quality of life. For example, is there a link between transport and access to leisure facilities that has a positive or negative effect on you?

d) Are there any influences on your quality of life where you live that are not yet on your spidergram? If so, add them now and develop them outwards.

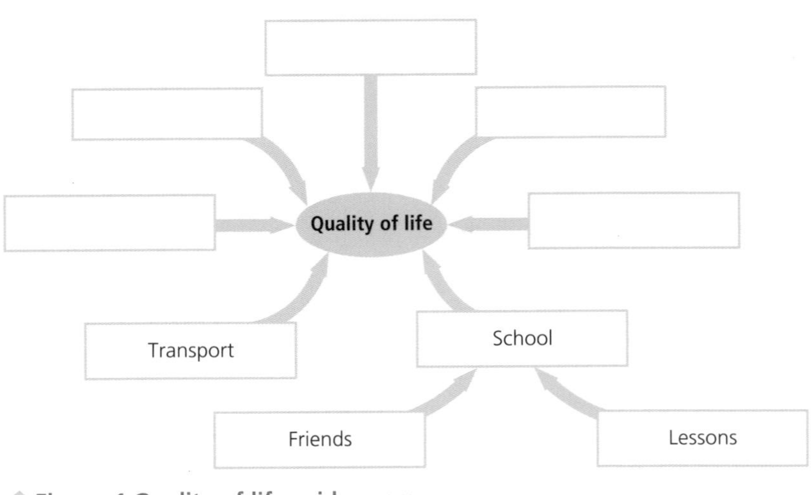

Figure 1 Quality of life spidergram

Figure 2 Influences on quality of life in rural Ghana

Knowing the basics

Revised

1 Study Figure 2.
  a) **Give one** feature *in the photograph* that could have a negative effect on the quality of life of these people.
  b) **Give one** feature *in the photograph* that could have a positive effect on the quality of life of these people.
  c) **Suggest** how each feature *shown around the photograph* may affect their quality of life.
  d) **Compare** your own quality of life with that of these people.

Exam tips

- Look carefully at the command words (highlighted in bold). Check their meanings on page 14. Only give what is asked for.
- Learn the key terms – for example, what is the difference between **rural** and **urban**? Your examiners expect you to know certain terms in the exams.

	UK	Ghana
Life expectancy at birth (years)	80	64
GDP per capita (wealth) $US	36,256	1,319
Adult literacy % (15 years+)	99	67

Source: World DataBank, 2013

Figure 3 Indications of development, UK and Ghana 2010

Rural – an area of countryside

Urban – a built up area like a town or city

Knowing the basics

Revised

1 Choose two of the indicators in Figure 3.
  Explain how each of these shows that people living in Ghana are likely to have a poorer quality of life than people living in the UK.

2 Which of the definitions at the top of page 18 refers to quality of life and which refers to standard of living?

Access to housing

How access to housing differs

Revised

Where do people live? There does seem to be an obvious answer to this question. Think about the area around your school. What differences are there in housing types? Are there groupings of similar types and sizes of houses or are they mixed up? Who owns the houses?

The legal and financial arrangements by which people live in their housing is known as **housing tenure**. It is quite varied.

1 Some housing is owned outright by the people who live in it, or they are paying off a loan called a mortgage from a bank or building society. This housing is said to be '**owner occupied**'.

2 In some housing the occupier pays rent to a private individual who owns the housing. This is called '**privately rented**'.

3 Existing housing is sometimes occupied illegally or houses are sometimes built illegally on land that is not owned by the builder/occupier. This is '**squatter' housing**.

4 There are two main areas of **social housing**:

- Some are owned by the local authority and are loaned to the occupier for a payment of money called rent. This is '**council rented**' or 'public' property.
- '**Housing associations**' are groups of people who own property, often apartments or flats. The occupier, who is a member of the housing association, pays rent to the association but also shares in the profit made by the association. 'Housing co-operatives' are one form of housing association in which people with shared values live together in a large converted building.

Since 2005 there has been a slow decline in owner occupied and social housing, and a steady rise of privately rented properties.

How access to housing affects people

Revised

There are advantages and disadvantages of each type of housing tenure.

Knowing the basics

Revised

1 a) Complete the following 'heads and tails' sentences to show some negative effects on quality of life of renting your house from a private landlord. The first one has been completed for you.

Heads	Tails
(1) May wait a long time for repairs	(a) ***so*** cannot improve living conditions.
(2) The property is never yours	(b) ***so*** may not be weatherproof for a while.
(3) Could have to leave at short notice	(c) ***so*** less money available for luxuries.
(4) Cannot alter it without permission	(d) ***so*** could end up homeless.
(5) May not pay a fair rent	(e) ***so*** will never stop paying for housing.

b) Suggest why each of the following may rent from a private landlord.

- A university student
- A recent immigrant to this country
- A newly divorced person
- A person who has frequent career moves

Exam tip

Each of the 'tails' statements on the previous page in the exercise above starts with 'so'. It is an *elaboration* of the original statement of disadvantage. Your examiners know it as a 'so what?' statement. Always give a 'so what?' statement when an exam question asks for *one* reason and offers you 2 marks for doing so. The first mark is for the simple 'reason' statement and the second mark is for its elaboration.

Stretch and challenge

Look back over your notes on this topic. Explain the advantages and disadvantages of:

- renting through a housing association
- obtaining a mortgage to buy your housing.

Revised

How access to housing is changing

Revised

Access to housing changes with time and is influenced by a number of factors. These include:

- **Borrowing a mortgage** – The possibility of a 'new buyer' being able to purchase a property changes as a result of several factors. The availability of 'affordable housing' is one such important factor. Even if there is housing that a person can afford to buy, the purchase will still depend on whether or not they can obtain a mortgage. This will, in turn, be controlled by how much money the lenders, the banks and building societies, are prepared to offer and on whether the people wanting the loan are able to save a large enough deposit. The deposit expected by a lender is usually about 10 per cent of the total cost of the property. It also depends upon the security of the borrower's job. It's not easy getting on the housing market!
- **Fewer young people entering the housing market** – This also has a number of effects on other people:
  - It makes it difficult to change to a larger house.
  - It reduces overall house prices.
  - It prevents sale of housing after the death of the owner.
- **Government policy** – Different national governments have different 'housing' policies. These change over time and have a large effect on the ability of people to buy or rent housing. Local governments have to work within the rules set up by their own national government. They do, though, have the responsibility to house homeless people.

Stretch and challenge

Look at the six features of the proposed Housing Bill below. Choose the two features you consider the most important in tackling the problem of inequality of access to housing for different groups of people. Explain why the features you have chosen are more important than the other four.

Revised

Exam tip

This 'stretch and challenge' exercise is similar to the final task in your problem solving exam. It is asking you to apply the knowledge and understanding learned in your lessons to a new situation. Make sure you consider both the advantages and disadvantages of your chosen features compared with those you consider less important.

A Housing Bill to improve lives and communities in Wales

Health, well-being, education, jobs and local communities – every part of life is affected by the homes people live in. That's why it's vital that everyone in Wales has a decent place to live ...

That's why we are drafting a new Housing Bill for Wales which will:

- place a new duty on local authorities to take all reasonable steps to find a suitable home for homeless people
- let local authorities use suitable accommodation in the private rented sector for homeless people
- introduce a compulsory licensing scheme for all private rented sector landlords ... to help improve standards across the housing sector
- give local authorities the power to charge more than the standard rate of council tax on homes which have been empty for longer than a year
- place a duty on local authorities to provide sites for gypsies and travellers where a clear need has been identified
- use more co-operative housing.

Source: adapted from a Welsh Government online bulletin, 7 December 2012

UK housing patterns

Different types of housing

Revised

Housing types include detached, semi-detached, terraced or town houses and apartments. They can range from those that are very large to those that are very small. Where these different types of housing are found is often quite varied and depends on how settlements have developed with time. Sometimes simple patterns can be seen, although some settlements like London are too big to show them.

These patterns can also give a clue as to their tenure in that, historically, different types of housing and their styles indicated whether the properties were owner occupied, privately rented or council rented. However, due to national government policies there has been a great reduction in council rented housing in the UK. It is quite possible that, for example, former high-rise council flats may now be privately owned, owned by housing associations, student accommodation, privately rented … or still council rented.

Different housing may also be suited to different groups of people. For example, the needs of a single person are quite different from those of a family with two children. People's housing needs also tend to change with time.

Formal settlements – homes where the householders have legal rights to the land

Informal settlements – homes where the householders have no legal rights to the land, i.e. they do not have legal housing tenure. Informal settlements are commonly known as shanty towns and squatter settlements.

	Central Business District (CBD)	Inner City	Inner Suburbs	Outer Suburbs
Cost of Living	High	Low	Medium	High/Low
Housing Types	Apartments	Old terraced blocks 1960s high rise flats and low rise houses Modern town houses	Mixed detached and semi-detached houses	Large detached houses, 1960s high rise flats and low rise houses
Ownership	Owner occupied Privately rented	Owner occupied Council rented Privately rented	Owner occupied	Owner occupied Council rented

↑ **Figure 4 A possible pattern of housing types in a large town or city**

Knowing the basics

1. Look at the cross-section of housing types in Figure 4. Describe how the housing type changes from the central business district (CBD) to the outer suburbs.
2. Draw a similar section to show changes in your nearest town or city.
3. How do the two sections compare?
4. Use information from Figure 4 to help you explain your own section.

Revised

Stretch and challenge

To what extent would you get a different pattern if you drew a section in a different direction from the CBD? Suggest reasons for the similarities and differences you would get.

Housing in poorer countries

Housing patterns

Revised

The pattern of housing is just as complex in many poorer countries as it is in the UK. However, again there are some very general patterns that may be applied.

Centre			Suburbs
Central Business District	High quality	Low quality	Squatter

- There is as wide a range of quality of housing as in the UK.
- A smaller proportion of people live in high quality houses and apartments.
- A greater proportion of people are either homeless and sleep on the streets or live in squatter settlements.
- While large squatter areas or towns exist it is also possible for isolated shacks or tented homes to be put up alongside high quality housing.

Features of the town or city that may disrupt the basic pattern from centre to **suburbs** include main roads, rivers, railways and industrial areas.

Squatter settlements

Revised

These **informal settlements** have a variety of names including favela, shanty town, spontaneous settlements. Whatever the name, they are unplanned and built on land that does not belong to the residents. They are built from any scrap materials that can be freely found.

Stretch and challenge

Revised

Suggest how each of the following factors is likely to influence where a squatter settlement will grow:

- Most people living in squatter settlements moved into the city from the countryside.
- Water is important for drinking, cooking and washing.
- It can be expensive to commute to work.
- The rich live near the city centres.
- Squatters have little access to private transport.
- Flood plain land tends to be avoided by **formal settlements**.

Stretch and challenge

Squatter settlements are often places of high resident unemployment, high crime rates and high disease rates. They are places that local police often refuse to enter. Use information from your studies to suggest how a local government might tackle the problems of its squatter settlements.

Revised

Squatter settlements usually lack basic amenities. Not having them can have negative effects on people living in the settlements.

Missing amenity	Effect on residents
A piped water supply	
Mains sewerage	
Street lighting	
Paved roads	
A source of electricity	

Knowing the basics

Complete the table on the left to show how residents may be affected by living in a squatter settlement that doesn't have these basic amenities.

Revised

Images in geography

Images of urban areas

Revised

Figure 5a Informal housing: the Algarve, Portugal. Not all low quality settlement is outside of Europe.

Exam tip

You may be asked to label or annotate a map, photograph, graph or diagram in the exam. Be sure that you know the difference between the two terms.

- '**Labelling**' is just stating what you see. For example, in this photograph you could label that it is 'next to roadside car parking'.
- '**Annotation**' is information added as explanation. In this case you are asked to annotate each feature to explain how it may affect quality of life. An annotation of the above label could be 'exhaust fumes lead to breathing difficulties'.

Knowing the basics

Revised

1. Label Figure 5a to show features of this housing that affect the quality of life of the people living in it.
2. Annotate each feature to show how these features affect the quality of life of people living in it.

Choose two different colours to show this information, one colour for your labels and the other for your annotation.

What's in an image?

You see images everywhere, for example, in newspapers, school textbooks and on television. They are mainly snapshots of a moment in time and are sometimes selected by the user to communicate a particular idea to the reader or viewer. That's why these three photographs are used here. In the background of Figure 5a are apartments that are more typical of this part of the Portuguese holiday coast. Those of Gabarone were chosen to show that many aspects of this 'Sub-Saharan Africa' urban area have much in common with what we are used to in the UK. In fact, there are many much poorer dwellings there, and some large houses as well that would equal the largest found in a typical British town.

In your problem solving paper you will be exposed to different viewpoints often backed up by images. Use them carefully when making your decisions.

Figure 5b Formal housing: Gabarone, the capital city of Botswana, Africa

Figure 5c Street scene, Gabarone

Knowing the basics

Briefly describe Figures 5b and 5c.

1. What feelings does each photograph give you about the place?
2. How might viewing images like these influence people living in a Botswana village similar to that shown in Figure 2 on page 19?

Revised

Stretch and challenge

1. Which two images on this page would you use to present a positive image of your own local area? Explain your choices.
2. Which two images would you most wish to avoid? Explain your choices.

Revised

Access to services

It's not just the house or housing area we live in that affects our lives. Our quality of life is also influenced by the services we have access to. One of the groups of services we use most days of every week are **retail services**.

How access to services differ and how it affects people

Revised

Historically there has been a definite pattern to shopping services in built-up areas. This has developed and changed over time as our **urban areas** have also grown and changed. The old pattern of terraced houses that grew in the **inner city** areas has been broken down and, as these have been replaced, many of the corner shops have disappeared. On the other hand, increased travel opportunities brought about by improved public transport services, and increased car ownership, have encouraged the growth of suburban supermarkets and huge out-of-town shopping centres.

Having services in a particular area is important. But far more important is whether people are able to easily access the services they wish to use, for shopping and entertainment, or in an emergency.

Catchment area – The area from which a shop attracts its customers.

Range – In this case, the furthest distance a customer is willing to travel to shop at a particular store.

Threshold population – The number of customers below which a shop will not make a big enough profit to stay open.

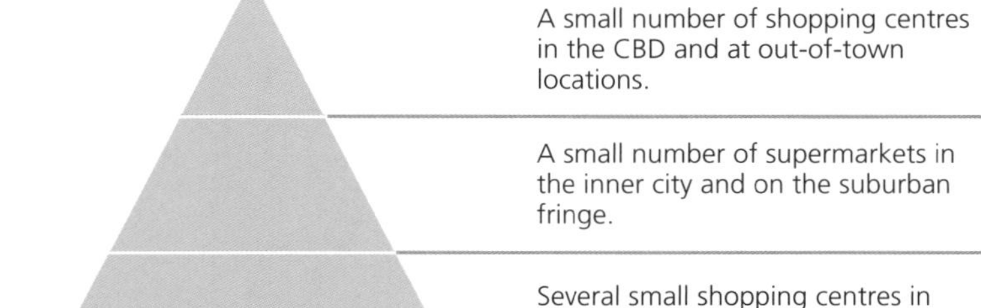

↑ **Figure 6 A shopping hierarchy**

Knowing the basics

Look at the section in Figure 4 on page 22. Use information from Figure 6 to add each of the types of shop to their correct place on the section.

Revised

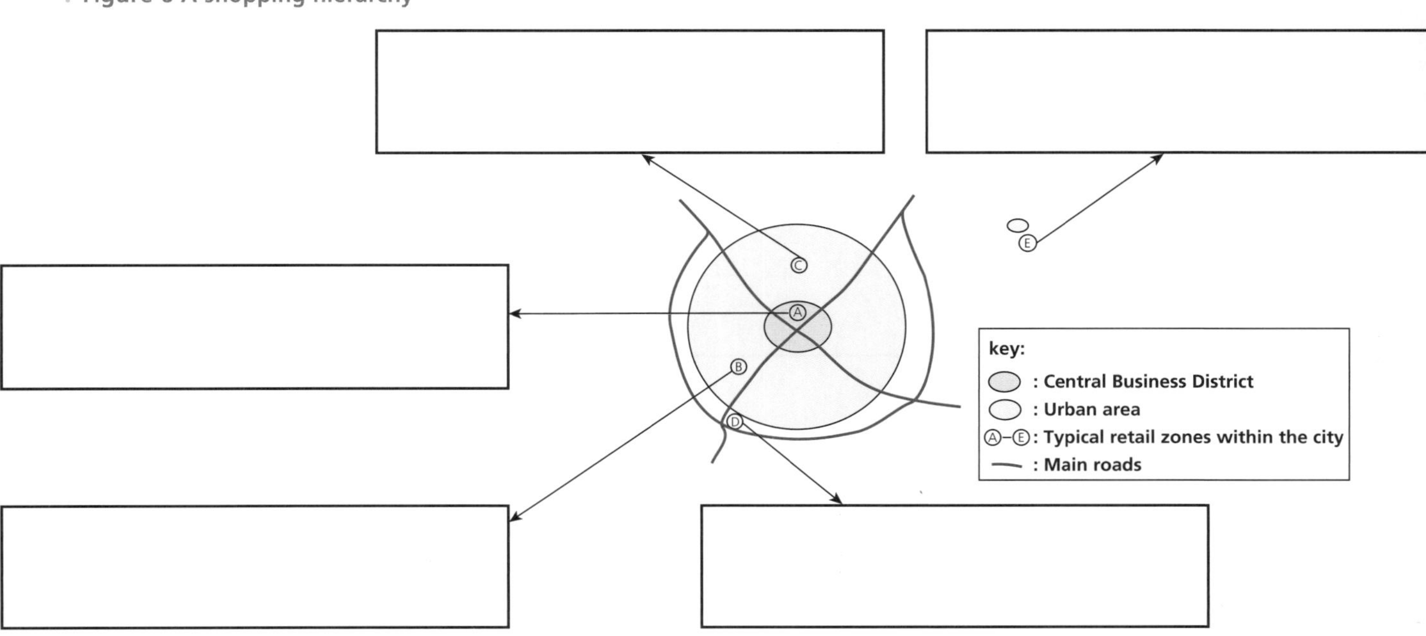

↑ **Figure 7 A typical pattern of retailing in UK cities**

Stretch and challenge

Revised

1 Complete these activities on Figure 7 on page 25 using information about the built-up area you have studied. For each empty box:

a) Write the names of each retail zone from your built-up area which corresponds to the position on the map.

b) Highlight each zone to show how well the shops in the zone are used:

Green = thriving
Amber = getting by
Red = struggling

c) In each box briefly explain why you feel each zone of shops is performing in the way you have shown.

How access to services is changing

Revised

High street shops in particular are now facing greater challenges than ever before. Many large shopping chains have closed down completely in recent years. These are the closures that make the news headlines, though there are many others where the company hasn't disappeared altogether but has closed down some of its shops that were unable to attract enough customers.

One of the main recent problems has been that of **economic recession**. As more people become unemployed or are not given pay rises that keep up with inflation, there is less money for people to spend on luxuries. They are said to have less disposable income (the money that is left after all the essentials of living have been paid). If there is less to spend, less money finds its way into shops which then struggle to survive.

Another threat to high street shops has come from the different shopping habits brought about by **changing technology** and some of these are highlighted in the table below.

Company	UK employees	UK stores	Troubles faced
Blockbuster UK	4,190	528	• Competition from internet firms streaming films as well as rentals via post
HMV	4,350	239 (230 HMV stores and 9 Fopp stores)	• CDs and DVDs sold more cheaply in supermarkets and online • Sales of other goods such as electrical gadgets and (now abandoned) expansion into live music didn't cover money lost from sales of CDs and DVDs
Jessops	1,534	187	• More competition from supermarkets and online retailers • Consumers often choose not to buy a dedicated camera due to improved quality of cameras on smart phones
Comet	6,611	236	• The recession led many to purchase high cost items like TVs less frequently • Sales of items like this are often made online, where consumers can find a better deal

Exam practice

Look at the table above.

a) What would be the total job loss if all these companies' shops closed? [1]

b) Explain one effect of the loss of these jobs on other retail companies. [2]

c) Give three reasons stated for the problems being faced by these companies. [3]

d) Suggest one advantage and one disadvantage to the customer of online shopping. [4]

Answers online

Online

How access to non-retail services varies between different groups of people

How do we use different services?

Revised

What other services do you use and how do they affect your quality of life? The frequency with which you use services can considerably vary. The ones that directly affect you on a daily or weekly basis include school and sport, cultural and recreation opportunities. Others are, perhaps, only noticed by you when you *need* to use them. These include such services as hospitals and doctors' surgeries.

Exam tip

Create a 'case study revision cards' for each of the 18 case studies you need for your exams. Always give specific detail about the actual place in your case studies.

Knowing the basics

Revised

Complete a 'case study revision card' in the spaces in the table below for the 'service other than retail' that you have studied in detail.

Accessibility (services) – the ease with which, in this case, a person is able to reach and use a service

Name of service:	**Built up area:**
Sketch map of distribution or location of service	Description of service facilities
Group of people: How accessible it is to them:	Factors that affect **accessibility** of this group: * * * * * *
Group of people: How accessible it is to them:	Factors that affect accessibility of this group: * * * * * *
Group of people: How accessible it is to them:	Factors that affect accessibility of this group: * * * * * *

Migrations

Types of migration

Revised

Migration is the movement of people from one place to another in order to live. This shouldn't be confused with commuting, which is the daily movement of people from their homes to their places of work.

Migrants fall into two broad groups:

- **Refugees** are people who move because they are forced away from the place where they live because their lives are in danger. This may be the result of a natural disaster or of a conflict, like a war. In the future it is possible that a separate group of refugees may be recognised, that of 'climate migrants', as extreme conditions force people to move away from their traditional home areas. Refugees have little choice of whether or not to move.
- **Economic migrants** move out of choice. They are usually attracted to a new place because it offers the prospect of a better job and better living conditions than in their original location.

The length of stay of either type of migration could be permanent or temporary, for example, when people move for a short time because of the effects of a major volcanic eruption. It may also be in order to take part in seasonal work. This is called **circular migration**.

Stretch and challenge

Complete the final two boxes in the activity below with two other migrations from your studies.

Revised

Knowing the basics

Revised

1 Complete each of boxes 1 to 4 to identify:

- whether the migration is of refugees or of economic migrants
- whether the migration is likely to be permanent or temporary.

You should realise by now that migration might be 'national', involving movement to and from different parts of the same country, or 'international', between different countries. Label each of the four boxes with an 'N' or an 'I' to show this information.

Type of migration: **Length of stay:** **1** The grape harvest in the vineyards of France attracts a large number of Spanish workers each year. They return to Spain after the harvest. The numbers involved depend on the unemployment rates in Spain.	**Type of migration:** **Length of stay:** **2** By December 2012 more than 135,000 Syrians had crossed the border into Turkey to escape the effects of the civil war in their country. Aid agencies have set up camps to shelter them.	**Type of migration:** **Length of stay:** **3** In the early twenty-first century many graduates from universities around India are attracted to the city Bangalore, or Bengaluru, by the high wages and pensions offered by the rapidly growing IT industry.
Type of migration: **Length of stay:** **4** In November 2010, nearly 280,000 people living on the fertile slopes of the Mount Merapi volcano, Indonesia, were forced into emergency shelters or crammed into transport to be with their friends and family who were farther away.	**Type of migration:** **Length of stay:**	**Type of migration:** **Length of stay:**

Rural to urban migration

Causes and effects of rural to urban migration

Revised

Migrations take place because of a number of push and pull factors. Push factors are those features of an area that encourage people to move away from that area. Pull factors are features of an area that *attract* people to that area.

1 The push of the countryside

In many of the world's poorer countries (see Figure 2, page 19) the migration is from rural to urban areas. Living conditions pushing people away from the countryside are likely to include factors such as:

- lack of work in area
- no school in the village
- a long walk to medical help
- an unreliable water supply

Knowing the basics

Revised

Write a sentence about each of the factors above to explain why it might help 'push' someone away from the countryside. For example, no school in the village **so** children can't read and write **so** they don't get well paid jobs.

Exam tip

Remember the importance of these 'so what?' statements. If you had been asked in the exam to explain one reason why people are pushed from the countryside, the first part of the sentence would have given you your first mark and the connecting 'so' part would be your second mark.

2 A response to disasters

Although living conditions may be very difficult in some areas, the final decision to migrate is often made when a disaster strikes, for example a flood or a drought. Some of the most devastating disasters are the result of problems brought about by people.

Zimbabwe

Millions of migrants crossed the border from Zimbabwe into Limpopo Province of South Africa in the early 2000s. For several years the country has been poorly managed, with high unemployment and rapid inflation making it difficult for people to live there. Some of the migrants are official **asylum seekers** or economic migrants with work permits. Most, though, are informal migrants who have no documents. They have moved mainly through desperation but also with the aim of sending money back to help family who remain in Zimbabwe.

Asylum seeker – a form of refugee with a well-founded fear of persecution in their country of origin for reasons of political opinion, religion, ethnicity, race/nationality, or membership of a particular social group

The migrants moved into a very poor province of South Africa. The average Limpopo household has less than 1000 Rand (the unit of currency) per month while households in the major city, Johannesburg, have 7175 Rand per month. Around 60 per cent of people in Limpopo live below the poverty line while only 20 per cent of people do so in Guateng, the province in which Johannesburg is found.

Looking at Limpopo's borders, perhaps you can now see why Limpopo has received refugees in recent years while also losing many people through migration to other South African provinces.

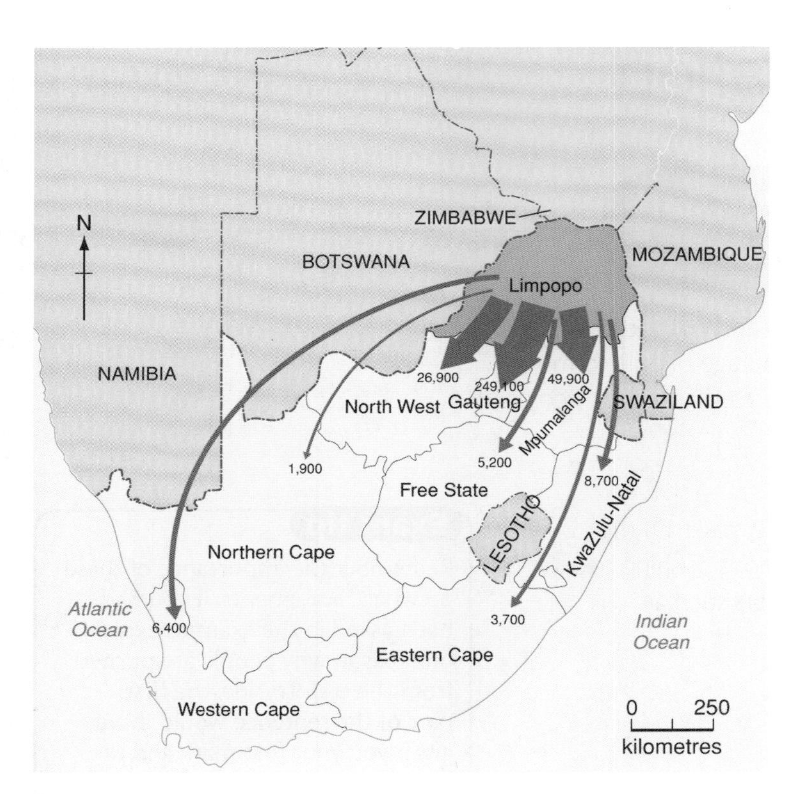

Figure 8 Migration from Limpopo province during 2005

Stretch and challenge

Revised

1. Read the section about Zimbabwe on page 29.
2. What is the difference between an asylum seeker and an economic migrant?
3. Underline in the paragraph two **push factors** that operated on these migrants. Explain how each may have 'pushed' them out of the country.
4. Three different types of migrant are mentioned in the paragraph. Underline each of these and explain how each may create difficulties for the South African government.

Exam tip

A question like the one below that asks you to use a map to describe patterns is quite common on the exam paper. When answering this type of question make sure that you:

- describe the general pattern
- use accurate figures.

Make sure that you do not offer explanations for the pattern you have described.

Exam practice

Study Figure 8 above.

a) Which province received the largest number of migrants from Limpopo in 2005? [1]

b) Describe the pattern of migration from Limpopo in 2005. [3]

Answers online

Online

Effects on the rural area

Revised

The migration of large numbers of people from the countryside to built-up areas like towns and cities has an impact on both the migrants' **source area** (the area they come from) and the **receiving area** (the area they go to). It is important to realise that the migrations being discussed now are **'free' migrations**, that is, these people move of their own free will as opposed to being pushed out by natural or human disasters.

(1) Migrants send money home
(2) Fewer people live in village
(3) Migrants are mainly young
(4) Mostly skilled workers leave
(5) Ideas are sent from city to village
(6) Migration is mainly of men

(a) Provides support for local businesses
(b) Some village jobs are incapable of being done
(c) A large dependent population is formed
(d) Takes pressure off food and water supplies
(e) Provides better seed and more cattle
(f) Families are split up

Figure 9 Effects of migration away from rural areas

Knowing the basics

1 Look at Figure 9 above. Each statement with a blue background is a feature of rural to urban migration in poorer areas of countries in Africa and Asia. The statements with a yellow background are possible effects of each feature.

a) Link each of the five pairs of statements.

b) Is outward migration a good or bad thing for a village? Why do you think this?

Revised

Effects of migration on children left at home

Social conditions in many African countries are changing rapidly. Many heads of households are migrating away from the rural areas to the cities. The migrants are often young adults. The cities they migrate to often have populations of which more than 25 per cent have HIV. Inevitably, this enables the spread of the disease to the villages when they return. This leaves many adults incapable of looking after their families.

The bottom line is that many households are now being run by the eldest child. It is almost impossible for these 'heads of households' to gain the education they so desperately need. They can't go to school when their time is mainly taken up with the adult responsibility of feeding their younger brothers or sisters. They are being forced to grow up too quickly.

Those whose parents have migrated live in the hope that they will return richer than they left or that they will send home regular payments to keep the family in food and shelter. Sadly, where HIV is involved, this doesn't happen very often.

Stretch and challenge

1 Read the news article on the left.

a) List ways in which the migration may affect the eldest children in some village families. You may wish to highlight them in the article.

b) Suggest why this family situation is much rarer in richer countries like the UK.

Revised

Urbanisation

Revised

Urbanisation is the growth of population and the physical size of larger towns and cities. Such growth may partly be the result of natural increase (where the birth rate is higher than the death rate), but it is mainly because of inward migration of people from rural areas. The proportion of the world's people living in urban areas increased from 30 per cent in 1950 to 47 per cent in 2000 and is expected to reach 60 per cent by 2030.

The pull of the cities

Revised

People are pulled to the cities by the perception of a better life. They expect to find all those features missing from their lives in the countryside … and they are also attracted by the positive reports on city life from people who have left the village before them. The idea that around 80 per cent of Johannesburg residents have an electric stove and fridge in their homes, and almost 90 per cent own a television, can be very persuasive in a village with no electricity supply. Add to that the prospect of a better paid job and access to education and health services, and the move to the city is a very attractive idea.

The reality

Migrants rarely find the housing or the job they are seeking. Much employment is informal and housing is likely to be a slum dwelling for those who are lucky. Many migrants sleep on the streets.

The reality of rural to urban migration in poorer countries is rarely as good as the migrants' expectations, either about the journey or what will happen once they reach their final destination.

↑ **Figure 10 A young lone migrant**

Knowing the basics

Revised

1. Look at the photograph of the young migrant in Figure 10. Life may be dangerous for migrants like this.
   a) Make a list of evidence from the photograph and captions to back up this statement.
   b) Add an elaboration (a 'so what?' statement) to help explain each piece of evidence.

Pressure on the city authorities

A large influx of people into an urban area is likely to put a great deal of demand on the town or city authorities. They will need to react to demand for jobs, education, health care and housing.

In many cities the authorities have worked with squatters to provide them with a brick-built, one- or two-roomed house, a piped water supply, mains sewerage, an electricity supply and a small garden. These are also provided with paved roads and street lighting. Such communities are known as 'site and service' schemes.

Knowing the basics

1. Draw a sketch of a brick-built house and label it with the features of a 'site and service' dwelling.
2. Explain how each feature is likely to provide a better quality of life for the residents than the squatter dwelling they previously lived in.

Revised

An uphill task

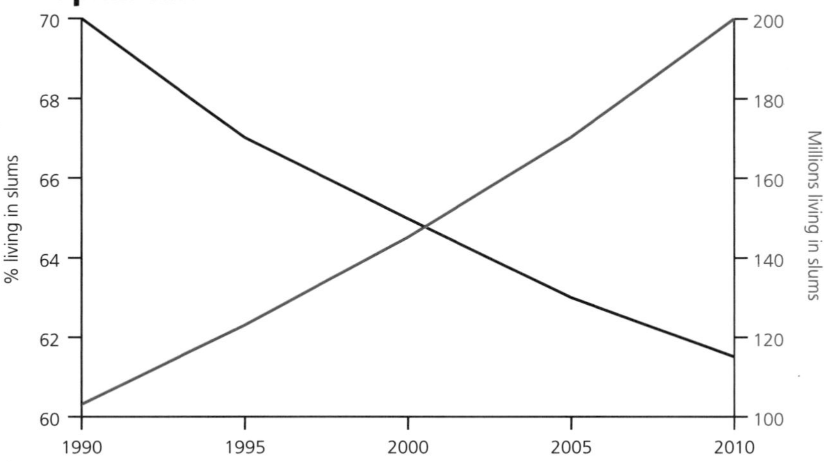

Figure 11 Trends in urban slum populations: Sub-Saharan Africa, 1990– 2010. Source: UN, State of the world's cities, 2010/11

Exam practice

Look at Figure 11 above.

a) Describe the percentage trend in the urban slum population of Sub-Saharan Africa between 1990 and 2010. Use figures in your answer. [3]

b) How does the trend in numbers of slum dwellers differ from this? [2]

c) Explain how the city authorities might tackle the problems they face. [5]

Answers online

Online

Exam tip

Look at part a) of the Exam practice question. It asks you to use figures in your answer. This means that you will not gain full marks if you don't do this. Giving a starting and finishing percentage would be enough when answering this question.

Planning issues in built environments

Who plans the changing built environment?

Revised

Who really makes the decisions that affect our lives? How much say in changes to your area do you or your parents have? To what extent does the real power rest in the hands of a small number of people? Can you really influence their decisions?

The main players in an urban planning decision are shown in Figure 12.

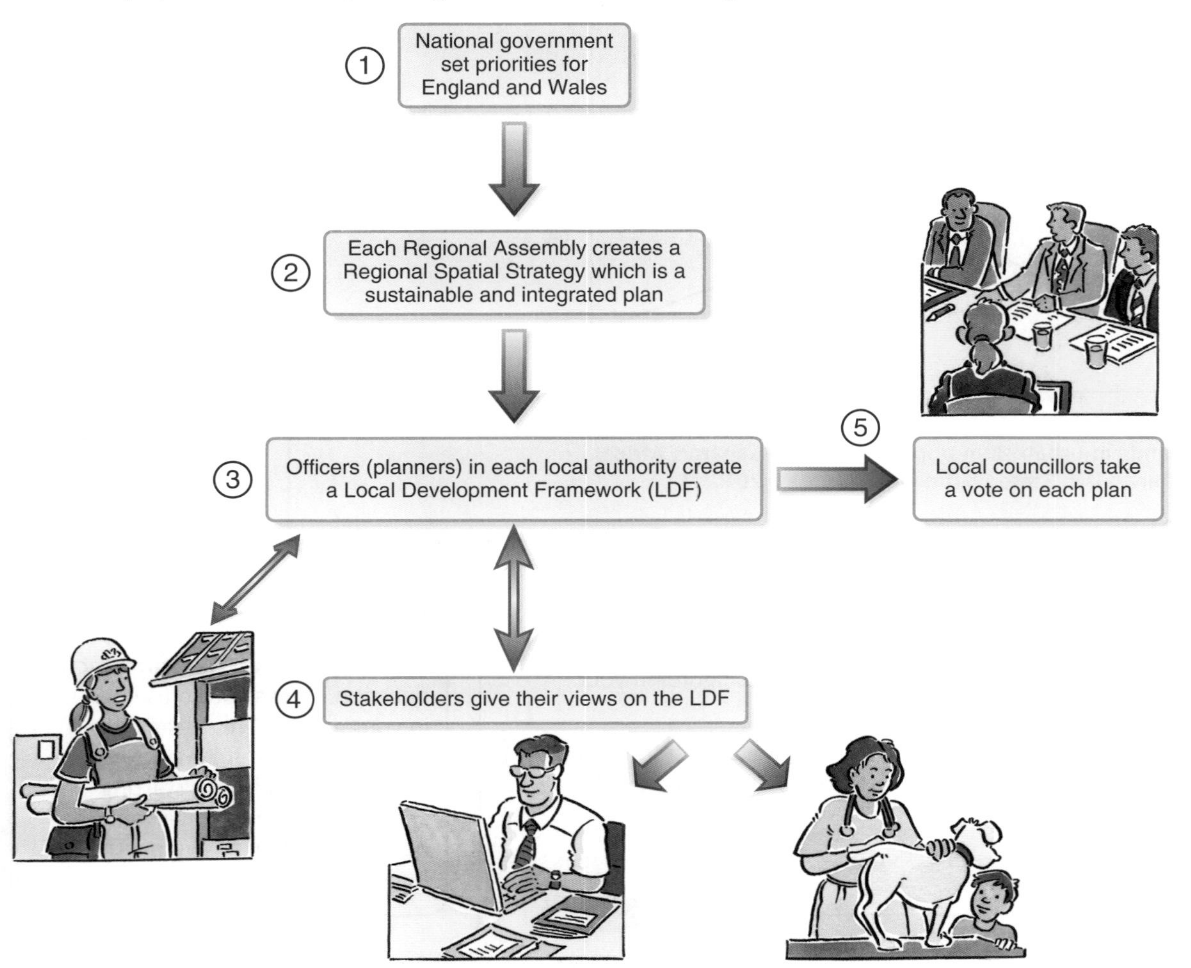

Figure 12 The three tiers of decision-making in the planning process

Knowing the basics

Revised

1 Describe the traditional planning model.
2 Who appears to have the most power? Explain your choice.

Exam tip

You should by now have completed the table on page 9. One of the case studies you need to know is 'One planning issue. The plans, stakeholders and reasons for conflict.' Make a 'case study revision card' for this case study using a similar structure to the table on page 27.

Influences on planning decisions

Revised

The Local Development Framework (LDF) is responsible for all aspects of planning in urban areas of England and Wales. Its work intends to create a better environment for all involved in the area, whether they live or work there or merely visit it.

The LDF makes decisions about such diverse aspects of the area as shopping, leisure, waste disposal, housing, work, crime, transport and environment issues.

Some major influences on the LDF are shown below.

Re-urbanisation

This involves a move of people inwards, towards city centres. They are attracted by the proximity of offices and shops in the centres at a time when both road congestion and high fuel prices make commuting expensive.

Gentrification

This involves the conversion and upgrading of existing buildings as an alternative to demolishing and replacing them with new properties, usually in inner city areas and the central business district (CBD). For example, large Victorian houses and old factories and warehouses may be converted to apartments. Gentrification often results in poor residents of a neighbourhood being displaced. This often results in increased average incomes and decreased average family sizes.

Greenfield site

An area of land that has not been used before for building. It has advantages to the developer in that there are no costs involved in demolishing buildings and removing the rubble, but it could be expensive to link to services like electricity and water supplies. Building on greenfield sites can be a sensitive issue in built-up areas where there is already little open space.

Brownfield site

An area for redevelopment that has already been built upon. The older buildings would be demolished before the new development takes place. This is costly for the developer. However, these costs may be offset by the advantage of the site already being linked to essential services such as water and electricity.

Green belt land

This is a government policy which is used to prevent the spread of cities into the countryside, in which an area of land surrounding the urban area is protected from development. The idea started in 1935 to help stop the expansion of London but it was not until 1955 that specific green belt areas were set up around the major **conurbations**. These areas are no longer as powerfully defended as originally proposed. The relative ease or difficulty of developing the green belt is now dependent on changing national planning priorities.

Stakeholders

These are people or groups of people who have either a direct or indirect interest in any planning issue. Most proposed developments in both urban and rural areas affect people who think quite differently from each other. These often join to form groups who work together towards influencing others to support the proposed development or to oppose it.

Sustainable communities

These are communities designed to have a minimum impact on the environment. Large-scale attempts to set up such communities include eco-towns of up to 15,000 homes, while individual eco-housing developments may be set up in existing built-up areas. However, these attempts usually stretch further than just setting up housing and also try to incorporate developing sustainable services, for example, eco-friendly transport.

Exam tip

Knowing key terms is very important in your exams. For example, you will lose 1 mark if you do not know the definition when asked 'what is meant by a brownfield site?' You could lose up to 6 marks if, instead, the question was 'explain the advantages and disadvantages of developing brownfield sites'. Make sure your key terms are well revised!

Knowing the basics

In the text boxes, each definition is shown in italics. Ensure you know not only what the definition says but also what it actually means. Write each feature on a small piece of paper or card and its definition on another. Mix these up and attempt a matching exercise to check your knowledge.

Revised

Stretch and challenge

Suggest how each of the boxed features on this page may affect the work of an LDF like the one responsible for planning the urban area you have studied.

Revised

What conflicts can be caused in a local planning issue?

Revised

Extending the Nottingham tram network

The first phase of the Nottingham tram network opened on 9 March 2004 with the link between Hucknall and Nottingham city centre. Phase 2 will be completed by the end of 2014 with two new routes linking the city centre with Clifton and Chilwell.

Negatives and positives

A development like this creates a great deal of disruption during the building period. Digging up roads and laying track makes life difficult for commuters. Local services like electricity lines and sewerage pipes need to be re-sited. This causes problems for local residents.

Additionally, there are longer-term impacts on the area the tram will pass through. Some family homes have been demolished and many others have lost parts of their gardens. Some houses will permanently suffer from noise pollution and privacy intrusion. To pay for the work, many Nottingham businesses, including the University of Nottingham, must now pay a 'parking levy'. This is usually passed on to the workers as a parking charge.

It's not all bad news, though. Phase 2 will:

- provide access to work for around 55,000 employees
- serve 20 of the 30 largest employers in Greater Nottingham
- put nearly 30 per cent of the Greater Nottingham population within 800 metres of a tram stop
- integrate with the road network through Park & Ride sites
- improve integration with bus services and the rail network
- take a further 3 million car journeys off Nottingham's roads.

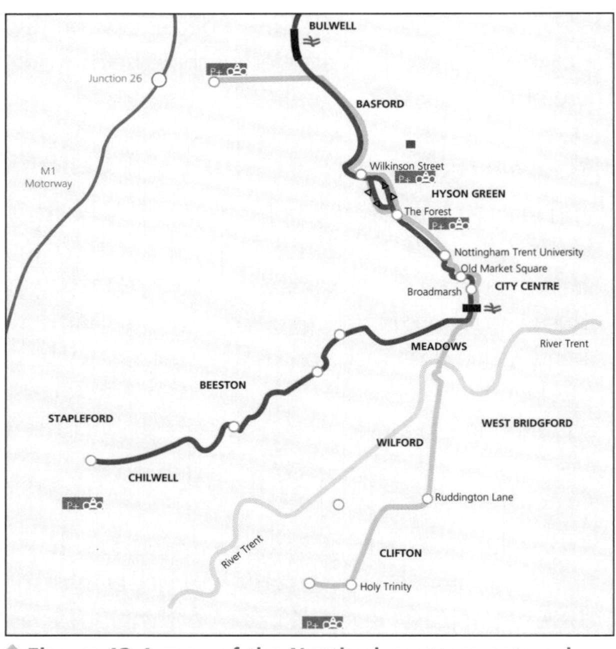

Figure 13 A map of the Nottingham tram network

Figure 14 Chilwell Road, Beeston: a planned closure for twelve months

Knowing the basics

Revised

1. Suggest two groups of stakeholders who would be against the development of Phase 2 of the tram network and two who would be in favour of it. Explain your choices.
2. How might the Nottingham tram development be regarded as a strategy for developing a sustainable community?

Choices

Some decisions about the route taken by the new tram lines were easy to make while others were much more difficult. The route through the centre of Beeston taken by the tram line to Chilwell is one of the more difficult ones. The developers put three different options into a 'consultation document' they sent to local people in early 2004.

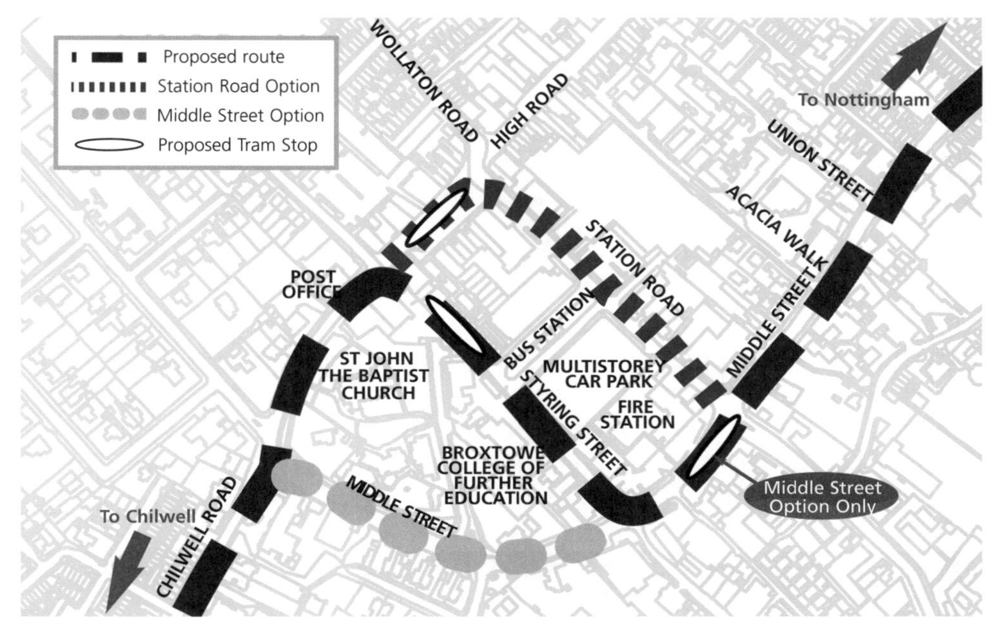

↑ Figure 15

	Station Road	Styring Street	Middle Street
Reliability	• Road often congested • Runs along major north–south route through town	• Mostly off major roads • New junction on Middle Street	• Middle Street often busy • Little impact on other traffic
Accessibility	• Tram stop in retail centre • Interchange with bus station	• Tram stop in retail centre • Interchange with bus station	• Away from town centre • Poor bus interchange
Costs	• Best • Low land purchase costs	• Middle • Higher land purchase costs	• Worst • High running costs with fewer passengers

Knowing the basics

Revised

The table above shows information about the possible routes of the tram through the area shown in Figure 15.

1 **a)** Highlight the features of each route that would help it to be chosen as the final route.
  b) Write a 'so what?' statement to explain the reason why that route would be chosen.

2 **a)** Highlight the features of each route that would discourage it from being chosen as the final route.
  b) Write a 'so what?' statement to explain the reason why that route would not be chosen.

3 Which route would you advise the local LDF to choose? Explain why your route is better than the other two options.

Exam tip

Task 3 is very similar to the final report you will have to write in your problem solving paper. Turn to page 122 of this book for ideas about how to structures your report.

Urban to rural migration

Causes and effects of urban to rural migration

Revised

A feature of some economically developed countries like the UK is the movement of people from urban to rural areas. In other words, people migrate from the cities to villages in the countryside. These people are also responding to a complex mixture of experiences of their lives in an urban area and perceptions as to how they might improve if they lived in the countryside. Such movements are resulting in **counter-urbanisation**, the growth of rural populations in areas that are accessible to the towns and cities by commuters.

↑ **Figure 16 Potential push and pull factors**

Some influences that encourage migration from urban to rural areas

These influences include perceptions of:

- air pollution
- fear of crime
- congested roads
- empty roads
- clean air
- friendly community.

Knowing the basics

Look at the six influences on the left on migration from urban to rural areas. Use them to help complete rows 2 and 3 of the table below.

Revised

	Urban push	Rural pull – perceptions	Explanation
1	Air pollution	Clean air	Concern about health effects of breathing air polluted by vehicles in city. Perception of much less impact in villages.
2			
3			
4			
5			

Stretch and challenge

Complete rows 4 and 5 of the table using push and pull factors of your own. Figure 16 may help you.

Revised

Exam tip

If you are asked to give two reasons why people migrate from one area to another do not give a matching pair. For example, don't quote 'air pollution' and 'clean air'.

Instead, give two completely different reasons, for example, 'air pollution' and 'fear of crime' in the cities.

Effects on rural areas and how they might be managed in a sustainable way

Revised

Villages fighting back?

People who migrate to rural areas pay high prices for houses, resulting in young villagers being unable to afford a house in the village and having to move away.

The newcomers also usually have different needs from those people who live there already. They often commute to work in nearby towns and cities and use the services they find there. Not only do house prices rise but there is also a sharp decline in village services such as post offices, banks, shops, schools, public transport and pubs.

Post office services

Café area

Local farm produce

Displays of local art and jewellery

Household essentials

Meeting room

Newspapers

Photocopying

Our Community-Owned Shop is Now Open!

www.plunkett.co.uk

Figure 17 Kirdford Village Stores, West Sussex is supported by the Plunkett Foundation, a charity set up to help maintain village services. It has two paid staff; the others are volunteers

PRESS RELEASE: New funding to revive rural communities

19 November 2012

New funding to help communities take control through enterprise

Rural communities across the UK are set to benefit from a new injection of funding to help them revive their villages through community enterprise.

The Plunkett Foundation has been awarded over £450,000 from long-term supporters the Esmée Fairbairn Foundation to create a new comprehensive support service for rural communities considering setting up or diversifying community-owned services – like shops and pubs – to help turn the tide on rural decline.

The funding, which follows the Foundation's hugely successful Village CORE programme, funded by Esmée Fairbairn, will provide a much-needed boost for communities considering community ownership, and will provide a combination of adviser support, training, feasibility grants, and opportunities for networking with other community enterprises. The support is specifically focused at the early stage of a community's ideas, and will help them progress to the next stage. Eligibility for support will depend on communities aiming to raise at least £10,000 themselves through community shares and support will be given to ensure communities are creating viable and sustainable businesses that engage the whole community.

James Alcock, Head of Frontline at the Plunkett Foundation, says: 'The Esmée Fairbairn Foundation has been a long-term supporter of Plunkett's work, funding our successful Village CORE programme which saw almost 100 rural communities being provided with specialist support and funding to save their village shop. Now, thanks to the continued support of Esmée Fairbairn, we are able to extend our work to other forms of enterprise, benefitting more communities across the whole of the UK.

'Our support through this programme depends upon the enterprises aiming to raise at least £10,000 by offering shares to the whole community. This approach has been extremely successful for enterprises like shops and pubs, allowing them to raise a significant amount of money in often a short space of time whilst ensuring the success of the enterprise is rooted in the whole community, empowering residents to make a difference and ensuring long-term sustainability.'

Source: adapted from a press release by the Plunkett Foundation

Knowing the basics

1 Read the newspaper article above.

a) How might Kirdford Village Stores help residents of the village?

b) How might it help local industries?

c) Why may it attract visitors to the village?

Revised

Stretch and challenge

Consider why village stores like this may have sustainable futures, free from external influence or from depending on external help.

Revised

Increased use of rural areas

Causes and effects of increased use of rural areas

Revised

People are taking retirement earlier than ever before, and workers also have more holidays than in the past. This all adds up to give much more leisure time. People can also move from place to place more speedily and cheaply than before. We can travel further to enjoy ourselves. Surely this can only be good news? Not always so (see Figure 18)!

↑ **Figure 18 Conflict in the countryside**

Knowing the basics

Revised

Look at Figure 18, a cartoon showing conflict in an area of natural beauty, then look at the table below. It has been completed to show that there is conflict between some farmers and hikers in the area. The cartoon shows that some hikers leave farm gates open and this allows farm animals to escape onto the road. So what? *So* the farm animals could be killed or injured by motor vehicles *so* costing the farmer a loss of earnings.

	Farmer	Hiker	Motorist	Picnicker		
Farmer		✓				
Hiker						
Motorist						
Picnicker						

1 Place more ticks on the table to identify other areas of conflict.
2 For each conflict you have identified, explain the reason for the conflict.

Stretch and challenge

Revised 

Add two other possible users of the valley to your table and explain how these people could add to the conflict.

Too many visitors?

Pressure on areas of outstanding natural beauty can not only cause conflict between the different users but can also damage the very environments we wish to visit.

Niagara Falls is a **honeypot site**. It attracts people in huge numbers to see and hear half a million gallons of water plummeting over the falls every second. About 6 million tourists a year come to see it. To cater for the tourists, the area has become filled with hotels and restaurants and many people earn a living from catering for the tourists.

↑ **Figure 19 Niagara Falls**

↑ **Figure 20 Tourist pressure at Niagara Falls**

Knowing the basics

Revised

Look at Figure 19 and then Figure 20, a sketch of the photograph.

1. Add labels to Figure 20 to show Niagara Falls, hotels, dual carriageway, café.
2. Annotate the figure to show how each of the features could attract visitors.

Honeypot site – a place of special interest that attracts many tourists and is often congested at peak times

Stretch and challenge

Revised

Suggest how the popularity of Niagara Falls may have made it a less attractive place to visit.

Exam tip

Look back at the Exam tips on page 24 for advice on labelling and annotation.

Creating and sustaining rural communities

Protecting areas of natural beauty

Revised

The pressure suffered by some areas of natural beauty has been long recognised. Special powers have been given to protect them from the activities of people. The most protected of these areas are called National Parks. The world's first National Park was created at Yellowstone in the USA in 1872. The first National Park of England and Wales was the Peak District National Park set up in 1951. The management of each National Park is the responsibility of its Planning Board.

The National Parks of England and Wales have three aims:

- to conserve and enhance the natural beauty, wildlife and cultural heritage of the National Park
- to promote opportunities for public enjoyment and understanding of the special qualities of the National Park
- to foster the economic and social well-being of communities living within the National Park.

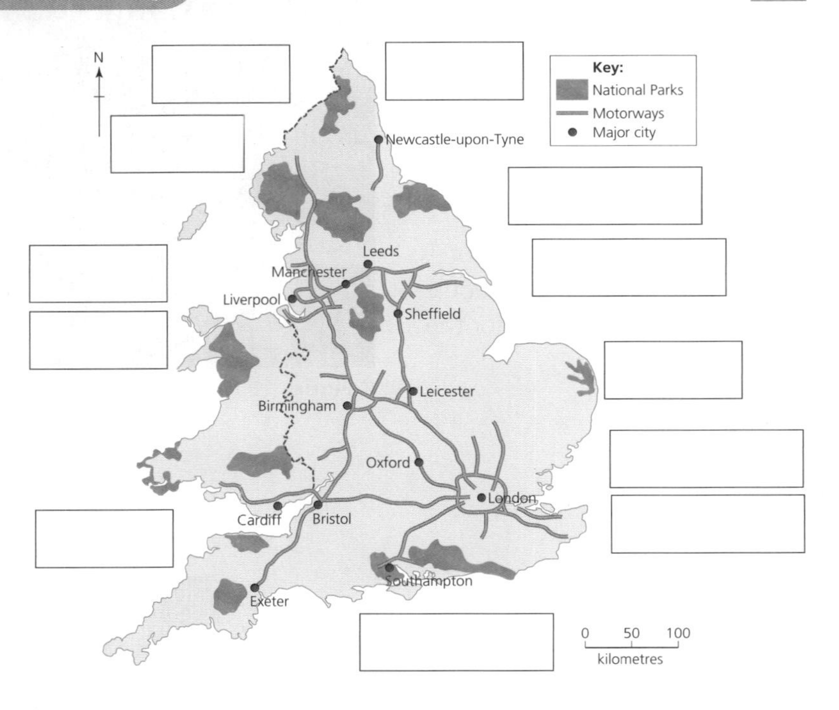

Figure 21 National Parks of England and Wales

Knowing the basics

Use an atlas to help you label Figure 21 to show the location of each National Park.

Revised

Stretch and challenge

The Peak District is the most visited National Park. Use evidence from Figure 21 to help explain why.

Revised

Knowing the basics

Revised

Read the three comments in the circles below. Each relates to one of the National Park aims.

1. Match each statement with the aim it relates to.
2. Suggest why the work of a National Park Planning Board is difficult.

Free talks on the geology of the National Park should be offered to visitors.

Areas should be closed to visitors to prevent any more footpath erosion.

A new quarry should be allowed to open inside the National Park.

Second homes: an issue facing National Parks

Revised

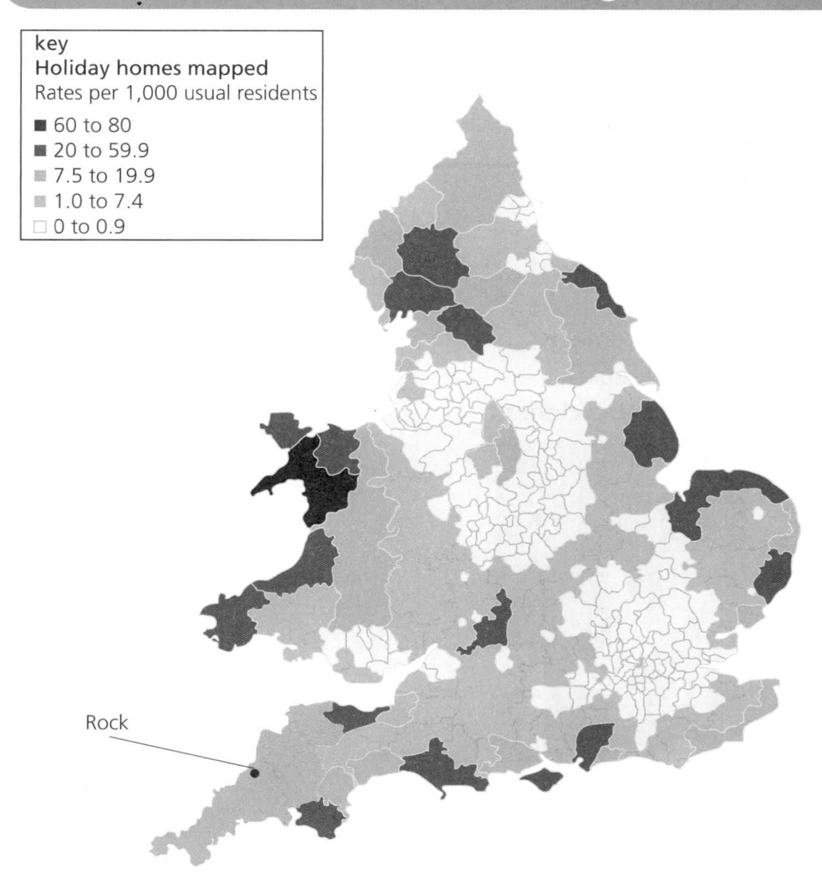

↑ **Figure 22 The percentage of homes in hamlets and isolated villages which are unoccupied, either because they are second homes or rented out as holiday homes**

Second homes at Rock, Cornwall

There is much for the visitor to enjoy: turquoise water, sandy beaches, good surf not too far away, a fine golf course, cafés, restaurants …

The shops are busy, the Blue Tomato café is full and builders are busying away on the few scraps of undeveloped land. (On the website of the local estate agent) … only three properties are advertised for less than £250,000. In contrast, there are 11 properties on the website priced at more than £1 million. (The same estate agent) also manages no fewer than 276 holiday cottages in and around Rock and Port Isaac.

With average wages in Cornwall only just scraping above £20,000, it is very difficult for people born here to stay. Cornwall council … last year called on the government to bring in new legislation to remove the council tax discount on second homes.

Source: adapted from the *Guardian*, 22 October 2012

Exam practice

1. **Compare** the distribution of holiday homes in England and Wales with the distribution of National Parks. [4]
2. **Suggest** and **explain** two reasons for the distributions you have described. [4]

Answers online

Online

Exam tip

When asked to 'compare' you must say how things are similar to and different from each other. You will help yourself do this by using words like 'similarly' when there is a positive correlation and 'whereas' when there is a negative correlation.

Knowing the basics

Revised

1. Read the newspaper article above.
2. What attracts people to the Rock area?
3. Explain the effects on the local area of having a large number of properties as second or holiday homes. Look at both advantages and disadvantages.

Stretch and challenge

Suggest what effects Cornwall County Council's proposed changes to council tax on second homes may have on the local area.

Revised

What weather and climate results from high and low pressure systems?

Depressions and anticyclones

Revised

A large amount of information is continually recorded at **weather** stations around the world. Data collected includes precipitation, temperature, wind speed and direction, sunshine and atmospheric (air) pressure. Air pressure is used to build up synoptic charts (weather maps) which show the positions and nature of low and high pressure areas. The centres of low pressure are called **depressions** and the centres of high pressure are called **anticyclones**.

Weather – Day-to-day changes in the atmosphere.

Climate – Average weather conditions over a period of usually at least 30 years.

Pressure system – A large mass of air having similar air pressure characteristics.

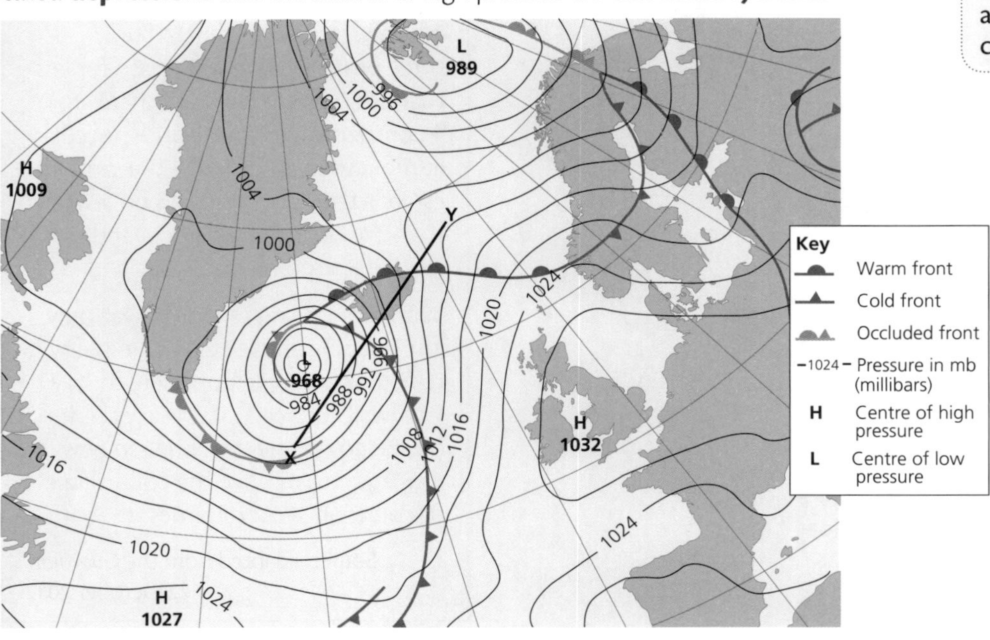

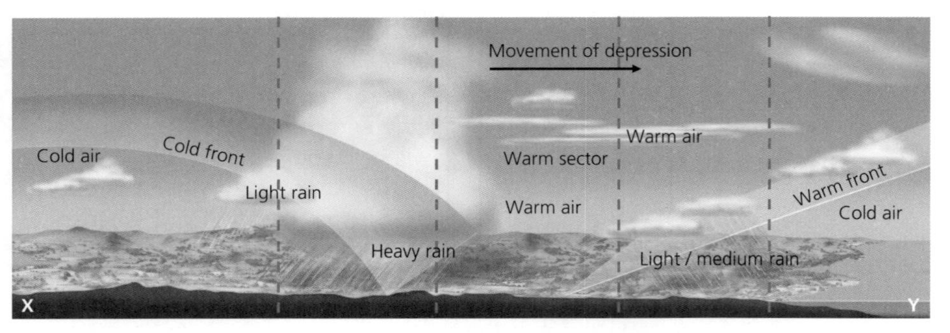

Figure 1 Weather map and section showing a deep area of low pressure in the north Atlantic Ocean on 4 September 2003

Feature	Depression	Anticyclone
Air pressure	Low: usually < 1000 mb	High: usually > 1020 mb
Vertical air movement	Rising	Falling
Wind strength	Strong	Weak
Wind circulation	Anti-clockwise	Clockwise

Figure 2 Main features of depressions and anticyclones

Exam tip

Do not confuse '**weather**' and '**climate**'. It is quite possible that a case study question might ask you to explore one or the other. When asked for a 'climate type' it would be rather foolish to write about the European heatwave of August 2003 or the effects of a severe depression like a hurricane. These are both weather events. Be careful!

Knowing the basics

Revised

Use information from the table in Figure 2 to help label the map in Figure 1.

1 **a** Add the following labels:
- depression
- anticyclone
- rising air
- falling air

b Add arrows to show the air circulation.

c Add a bold arrow to show the movement of the depression.

2 The section in Figure 1 shows the different weather that will be experienced as a depression passes over an area. Label the following text boxes from 1 to 5 in order to show the sequence of weather at Point Y as the depression passes over it.

Exam tip

In the exam you may be asked to label a map like this. Remember that labels only *show what is there*. You are not being asked to annotate so do not add any explanation.

(a)	(b)	(c)	(d)	(e)
Thick cloud	High cloud	Lowering cloud	High, light cloud	Cloud and sunny intervals
Heavy rain	Little or no rain	Rain	No rain	Showers
Cold	Cold	Cold	Warm	Cold

Types of rainfall

Revised

Rain falls in a depression as air rises. It is called **frontal rain**. Rising air also results in rainfall as air heats and rotates, and as air passes over hills and mountains (**orographic rain**) or in a thunderstorm (**convectional rain**). It does so because of the following sequence:

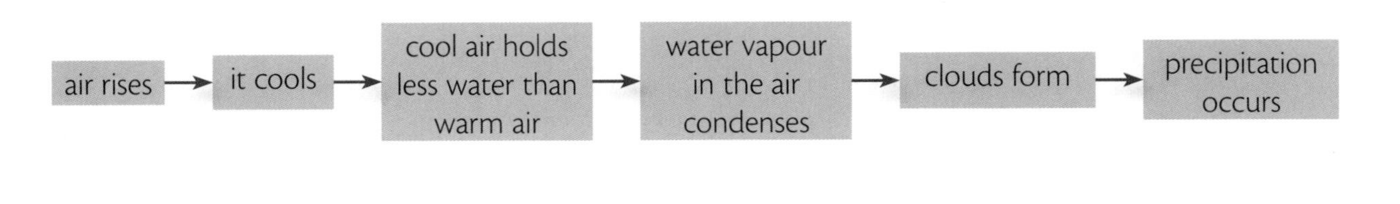

Knowing the basics

Revised

Complete a similar sequence for air sinking in an anticyclone.

Stretch and challenge

Revised

Explain why no rain fell over the UK on 4 September 2003.

How do anticyclones and depressions affect activities and quality of life?

How they affect everyday life

Revised

On page 45 you were asked to explain why no rain fell over the UK in early September 2003. Clearly it is because of the falling air in a depression which stabilises the air, discouraging cloud formation and rainfall. **How does this affect people?** As there will be clear skies, the air temperatures will rise quickly during the day. There is the possibility of weeks of hot and dry weather. These may bring positive impacts on quality of life in relation to, for example, outdoor sports (and ice cream and cold drink sales) and negative impacts on water supplies and healthcare services.

The effects of winter anticyclones are quite different. There is no cloud blanket to retain heat. The clear skies result in very cold conditions with severe frost and, often, the formation of thick fog. The effects on people could include the cancellation of outdoor sports due to frozen pitches, and hazardous travel conditions due to ice and fog. On the other hand, brighter days combat Seasonal Affective Disorder (SAD) and sales of hot drinks may increase.

We experience events like these, and the passages of depression, on a regular basis with little major disruption of our lives. It is when weather events become severe that real problems occur.

Severe weather events

Revised

A cyclone is a severe depression. Severe depressions are also called hurricanes. Cyclone Nargis hit southern Myanmar in early May 2008. The story of Nargis is shown below.

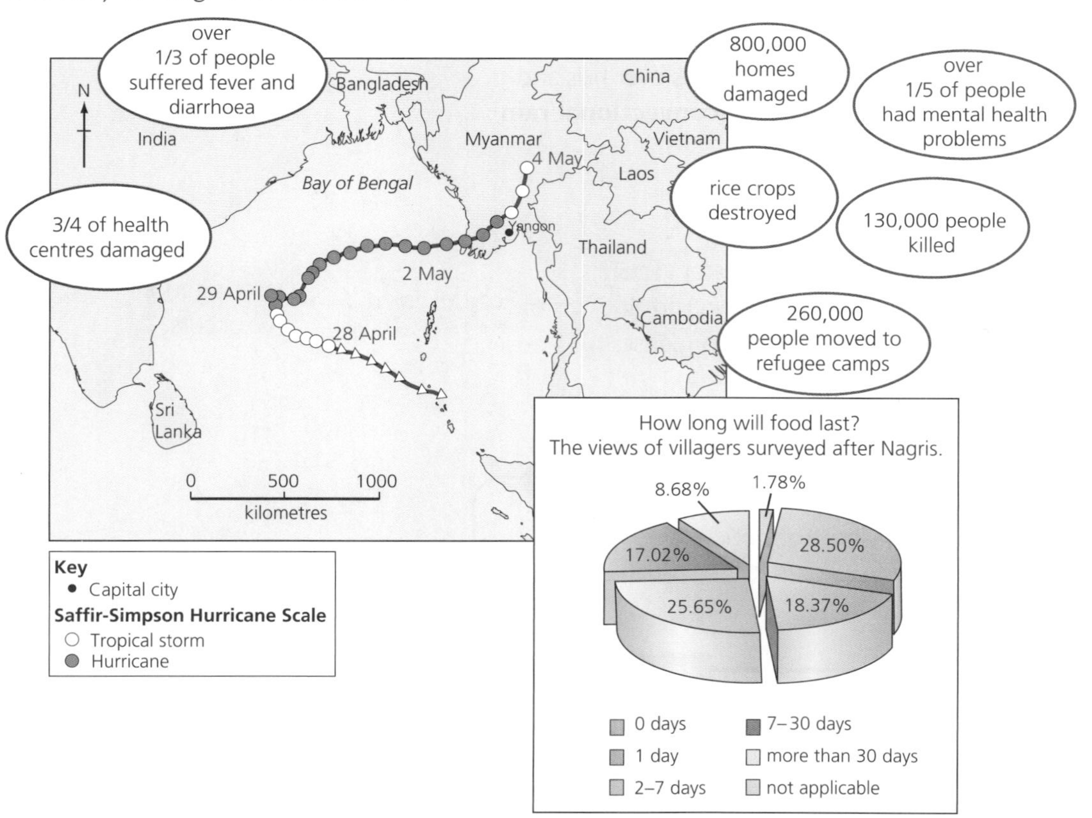

Figure 3 The story of Nargis

Exam practice

Use Figure 3 on page 46 to complete the following questions. The number of marks available for each question is shown in brackets.

1 Describe the path taken by Cyclone Nargis between 28 April and 4 May 2008. [3]

2 Suggest how the cyclone may have affected the lives of people in southern Myanmar. Use evidence from Figure 3 in your answer. [6]

Answers online

Online

Figure 4 Farmland in a winter anticyclone

Knowing the basics

Look at the photograph in Figure 4. It shows an area of farmland in the valley of the River Trent during a winter anticyclone.

1 Suggest ways in which these weather conditions might affect the everyday life of a farmer.

2 Compare this with ways in which the weather conditions would affect a typical day in your life.

Revised

Knowing the basics

Revised

Complete the table below to show the conditions brought to people in the UK by different types of weather events. Part of it has been done for you.

Weather event	Conditions	Positive impacts	Negative impacts
Depression	• Unsettled weather. • A period of cold rainy conditions, followed by a spell of drier, warm weather before heavy rainfall, a return to cold showery weather.		
Summer anticyclone			
Winter anticyclone			

How does drought affect people?

What is drought?

Revised

Drought is an extended period during which an area receives less rain than would normally be the case. In other words, it is not an actual figure in millimetres, but a relative one in which people and ecosystems have to cope with much less rain than they would normally receive: it is a difficult situation to be in!

How does drought affect people in richer countries?

Revised

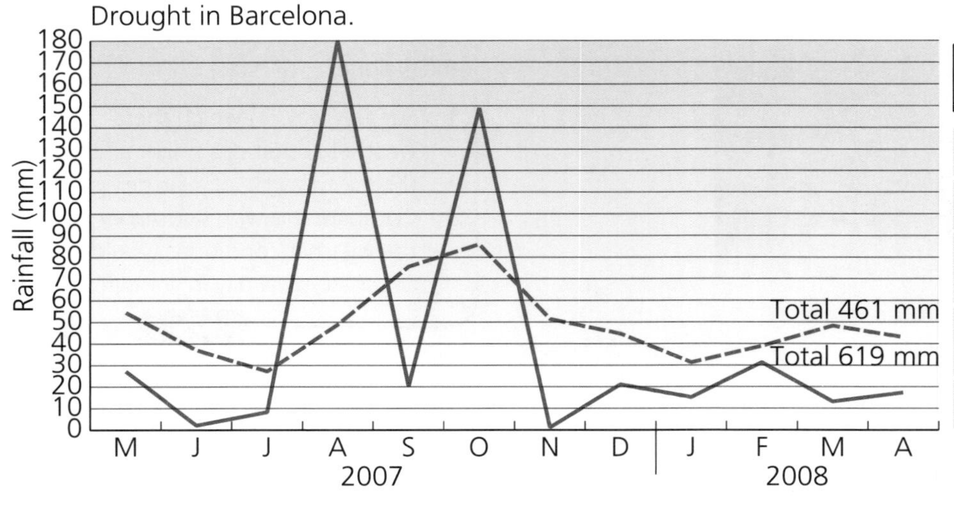

Figure 5a The cause of drought in Barcelona 2008

1 No watering of gardens
2 No car washing
3 10 per cent of public fountains turned off
4 No filling of private swimming pools
5 Fines for illegal water use

Figure 5b The official response

Exam tip

This exam practice question contains command words highlighted in green. Look back at page 14 to remind yourself about them.

Exam practice

Attempt one of the following sets of questions.

Higher Tier

a) **Describe** the pattern of precipitation between May 2007 and April 2008 in Figure 5. [3]

b) **Compare** this with the average precipitation. [3]

c) **Describe** how this precipitation difference affected the quality of life of people in Barcelona. [4]

Foundation Tier

a) **Complete** the following passage using Figure 5.

'Average precipitation varies between a low of 27 mm in __________ and a high of __________ mm in October. There is precipitation in all months. In contrast, the actual precipitation shows two months with none at all, July and __________. Differences between high and low are much greater with a maximum of precipitation __________ mm greater than average in October 2007. The actual precipitation was lower than average by __________ mm. [5]

b) **Describe** two ways in which people's quality of life was affected by the drought. [4]

Answers online

Online

What causes the droughts?

Revised

You should already know that stable air conditions result in little or no rain. World air mass patterns are slowly changing to bring different weather conditions. Longer periods of anticyclones over Spain will lead to a reduction in rainfall, potentially as much as 20 per cent less on average in the first half of this century (compared to 1950–99). At the same time more *frontal rain* is expected to fall on northern Europe.

Can water be supplied sustainably?

Revised

It is not just climatic change and rainfall variation that causes water supply problems. We have increased our demand for water, especially in richer countries. To meet this demand, water is sometimes transferred, often across borders, through pipes and canals from areas with a surplus to those with a deficit.

Knowing the basics

Revised

Look at Figure 6 which shows Barcelona's water supply.

1. Make two lists: one of solutions that will help provide a sustainable water supply in the future, and another of those that you feel are unsustainable.
2. For each solution, explain why you feel it is either sustainable or unsustainable.

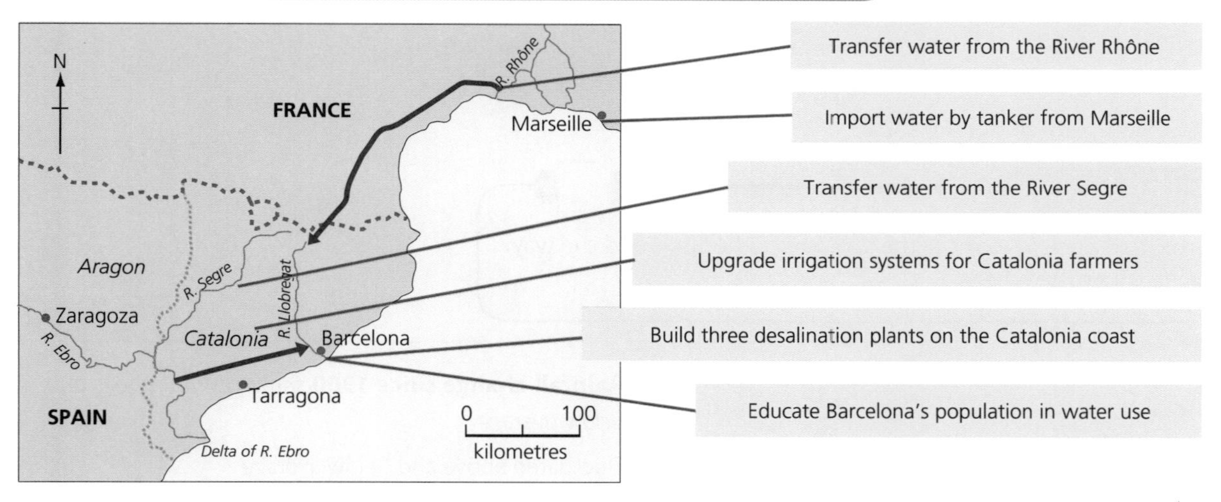

Figure 6 Meeting Barcelona's water needs

How big should a water scheme be?

Revised

Many governments have developed large-scale, multi-purpose water schemes. These usually involve damming a major river to create a huge lake.

Advantages	**Disadvantages**
Prevention of flooding	Drowning of settlements under the lake
Provision of water for farms and house	Reduction in natural supply of fertiliser to farmland
Generation of electricity	Reduction in water supply further downriver
Attraction of secondary industry	Destruction of land and river habitats
	Expensive to set up and operate

How does drought affect people in poorer countries?

Revised

The area bordering the southern edge of the Sahara desert contains some of the poorest countries in the world. The ways in which people living there respond to drought differ quite markedly from responses in the richer world.

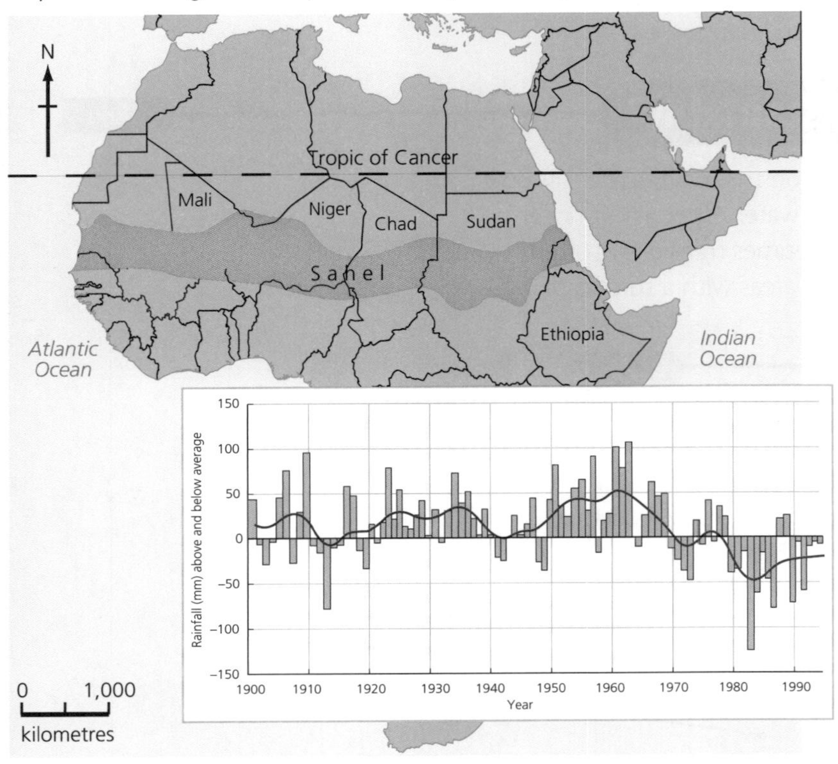

There is little water in the well. Grazing for my goats has dried out and I have lost many animals. Even wood for fires is scarce. Crops just die in the fields. Many of my friends have moved to the city.

My hut is on the edge of town and I take my turn at the shared tap. Water is only supplied for 8 hours a day and the queue is long. The water pressure is so low that it takes a long time to fill my large container. Dysentery and other water-borne diseases are common in this neighbourhood.

↑ **Figure 7 Continuing drought in the Sahel**

Knowing the basics

Revised

Complete the boxes below to show the location of the Sahel and the way in which rainfall fluctuates there.

Location

The continent of Africa

- Depth from north to south is:
  ____
- Length from east to west is:
  ____
- Countries affected are:
  ____
  ____
  ____
  ____
  ____
  ____
  ____
  ____

Rainfall change since 1900 *shown in mm above and below average*

Fluctuated above and below average

- Highest is: _______ mm above average
  Year: _______
- Lowest is: _______ mm below average
  Year: _______
- The rainfall trend from 1900 to 1995 was:
  ____
  ____
  ____
  ____
  ____
  ____
  ____

No water in well.
So what? Water dirty and rationed.
So diseases and dehydration.
So death.

Rural

Urban

Knowing the basics

Effects of recent changes on people

Read the quotations with Figure 7 on page 50. Copy and complete the diagrams on the left to help you organise effects of the drought on life in urban and rural areas. You may wish, instead, to attempt this for a different drought case study you know.

Revised

Intermediate technology: a sustainable option?

Revised

Large-scale water transfer or water storage schemes are not always the best ways to combat drought. In some countries governments and non-government organisations (NGOs) like WaterAid are setting up small-scale schemes in farming villages. These include the use of plate cisterns, underground dams, piledriver wells and rainwater harvesting. Such small-scale strategies are easy to install and cheap to operate and maintain.

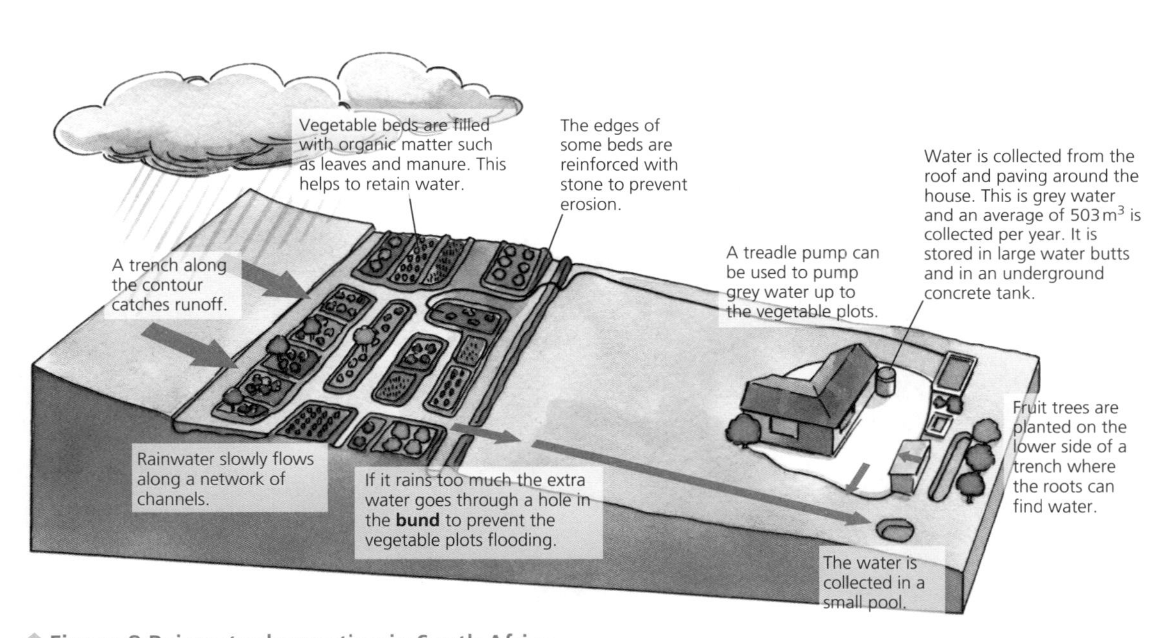

↑ **Figure 8 Rainwater harvesting in South Africa**

Knowing the basics

1. Make a list of three ways in which the rainwater harvesting scheme in Figure 8 gets maximum use out of the available water.
2. How might this scheme be considered sustainable for a small farm?

Revised

Stretch and challenge

Large-scale and multi-purpose vs. alternative technology

Which attempt to tackle drought do you prefer – the big scheme or the small scheme? Why do you prefer this? Who would disagree with you and who would agree? Why?

Revised

Climate differences and influences

Climate types

Revised

Geographers have divided the world into a number of different climate types. Although weather will vary within these large climate regions, the average rainfall and temperatures are similar in all parts of any one region.

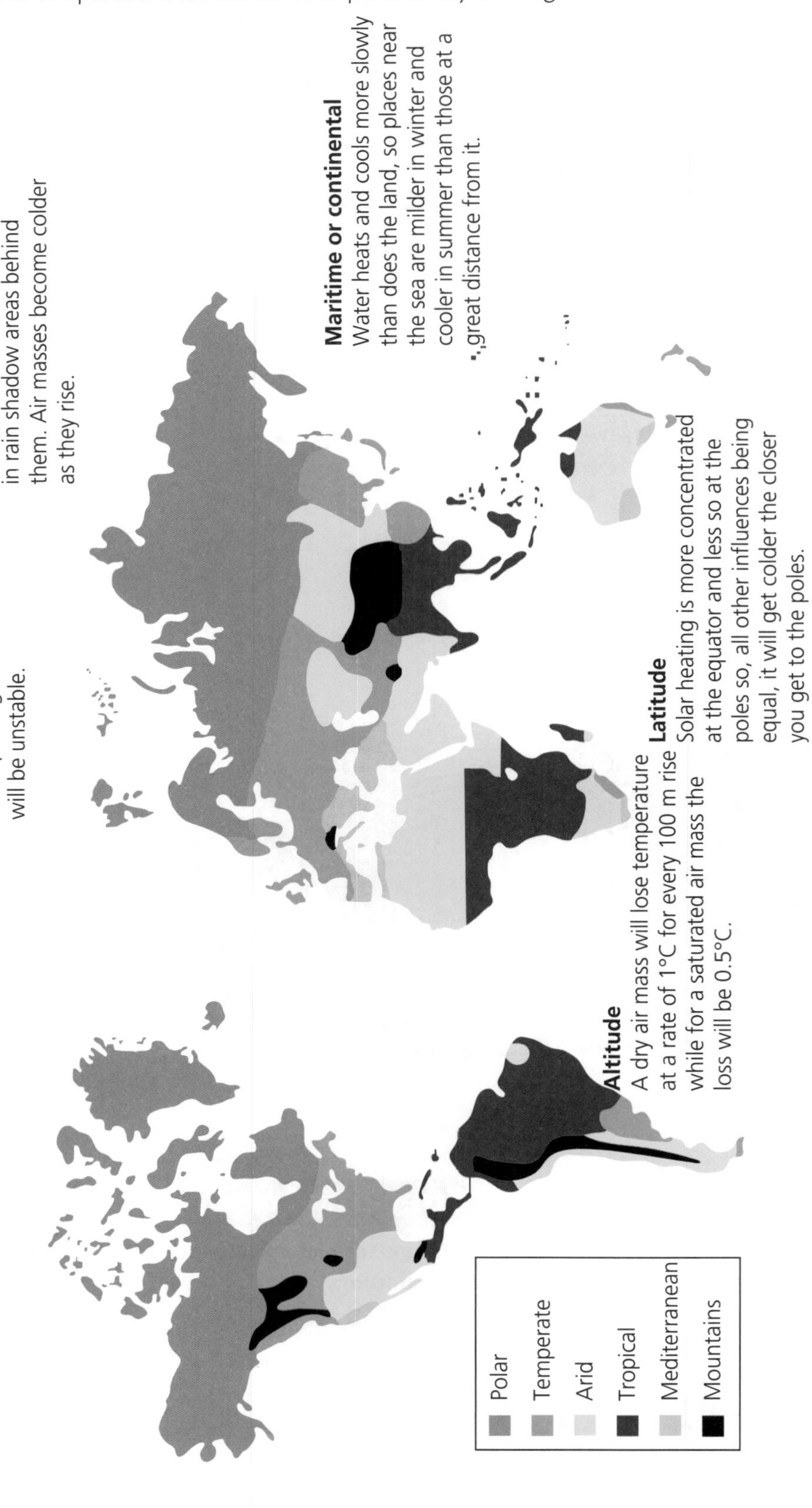

Figure 9 A simple division of the world into climate regions. Surrounding it are the major influences on world climate types.

Knowing the basics

Revised

1 Outline and label the European climate region and the tropical climate region you have studied on Figure 9.
2 Complete the table below for these two climate regions.
Use information from Figure 9 and the boxes below to help you.

	European climate **Name:**	**Tropical climate** **Name:**
Influences creating this climate type		
Main rainfall characteristics		
Main temperature characteristics		

How does climate affect ecosystems?

Revised

An **ecosystem** is a system of links between plants and animals and the habitats in which they live. Ecosystems include not only living (**organic**) parts but also **inorganic** elements such as the climate, and the particles of broken rock that combine with decomposing organic material to form soil. Large ecosystems are called **biomes**. The map in Figure 10 shows the distribution of one of the world's biomes.

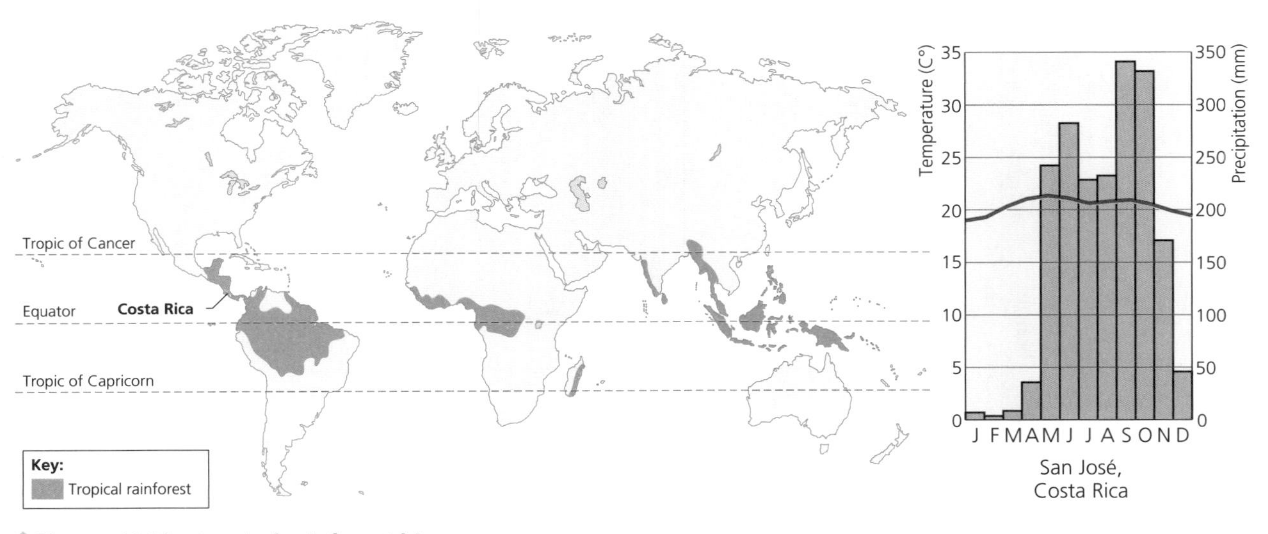

Figure 10 The tropical rainforest biome

The tropical rainforest relies on rapid recycling of dead plant material. This is possible in the rainforest climate. **Deforestation** exposes a fragile soil. This quickly erodes and the chances of forest regrowth are low.

Figure 11 Deforested land in Costa Rica

Deforestation – the cutting down or burning of trees

Exam tips

When answering this question, remember:

- to name areas. In this case you will need to use an atlas to help.
- to be specific. Which coast of India has an area of rainforest?
- to make general statements and add exceptions. Where are areas of rainforest found *outside* the tropics?

Exam practice

Describe the distribution of the world's rainforests. [3]

Answers online

Online

Knowing the basics

Revised

Some basic skills work … and a little understanding.

1 Label Figure 11 to show the following: uncleared rainforest, cleared rainforest, two other named pieces of evidence of human involvement in this area.

2 Land exposed in this way to wind and heavy rain is likely to suffer soil erosion.

a) Identify and label the area of the photo experiencing soil erosion.

b) Complete the exercise below to link the heads and tails to show some effects of soil erosion.

Heads	Tails
(1) Minerals in the soil are leached downwards	(a) so carrying capacity is lowered and floods are more frequent.
(2) Soil particles are washed into rivers	(b) so soil becomes less fertile.
(3) Soil material is deposited on the riverbed	(c) so river water is less clear and fish habitats are destroyed.

Exam practice

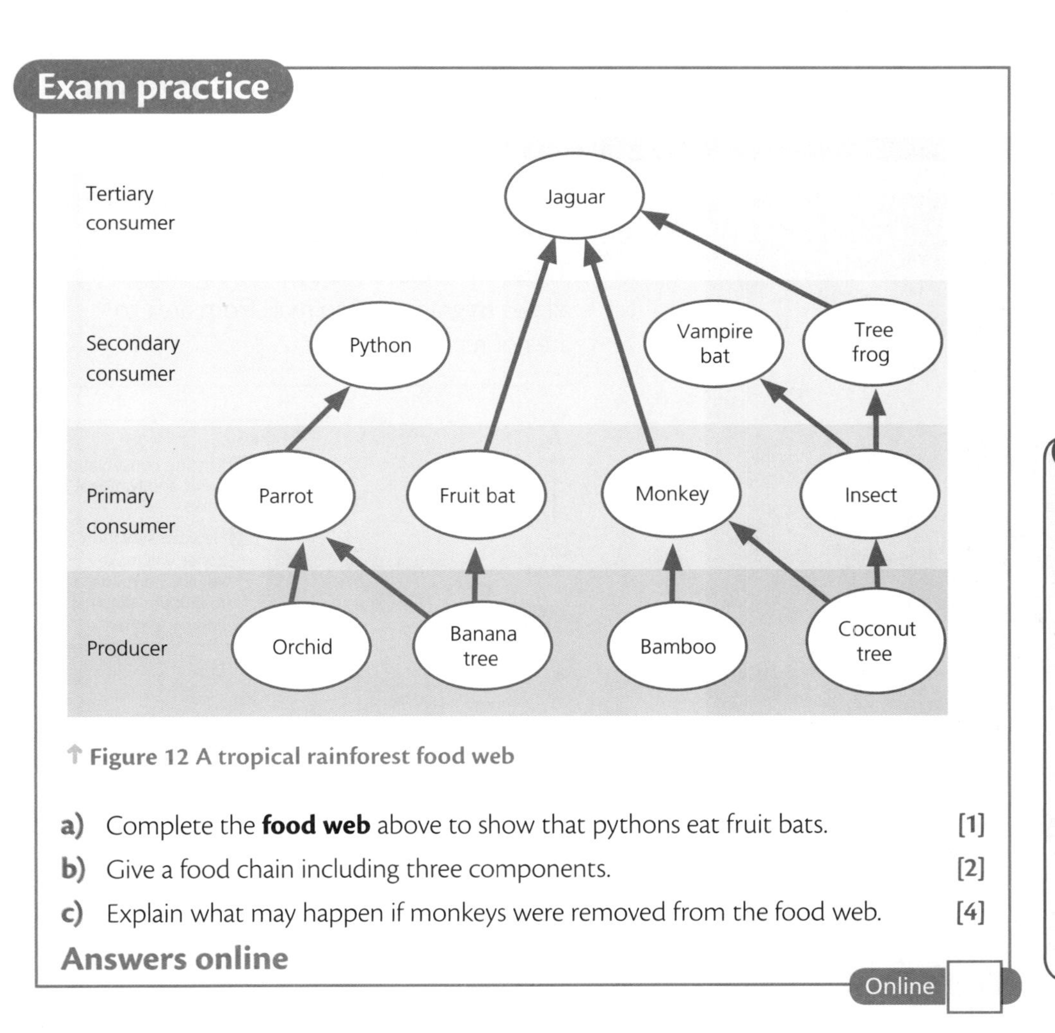

Figure 12 A tropical rainforest food web

a) Complete the **food web** above to show that pythons eat fruit bats. [1]

b) Give a food chain including three components. [2]

c) Explain what may happen if monkeys were removed from the food web. [4]

Answers online

Online

Exam tips

- In a) make sure your arrow is pointing towards the consumer.
- In b) you will need to list, in order, one producer, one primary consumer and one secondary consumer. Link them with arrows.
- In c) you need to have a sequence of effects, each a consequence of the previous one. Link them as a series of 'so what?' statements.

How might ecosystems be managed sustainably?

It is clear from pages 54–55 that there are many negative effects when people interfere with ecosystems. These effects may be deliberate, as in the case of rainforest destruction. They may be unintentional, as with the ongoing desertification of the Sahel, the southern edge of the Sahara desert; this is partly the result of global warming.

The destruction may be on a large scale, as you have already seen, or may be as small as using a local pond as a dump for waste.

Ecosystem management – Regulating the use of ecosystems so that human needs and those of the ecosystem are met.

What is meant by 'sustainable'?

Revised

This is the use of resources and environments in ways which allow them to continue to be used in the future, that is, not destroying them for short-term gain. Sustainable development is not restricted to natural environments. It equally applies to urban areas and economic development.

Sustainable development of an ecosystem: the tropical rainforest

Revised

Wildlife corridors

These are established by planting trees. This allows migration of animals from area to area of remaining forest.

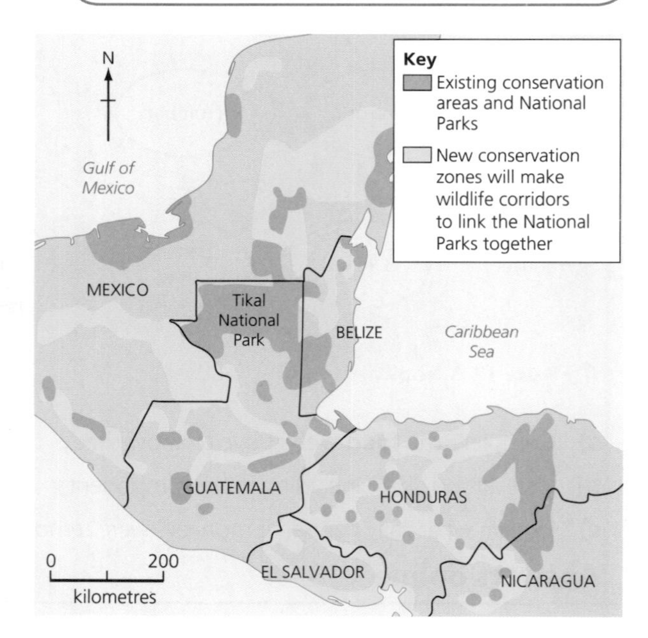

Figure 13 Protecting the Central American forests: wildlife corridors

Knowing the basics

Biodiversity is the variety of different plants and animals found in an area.

1 Study Figure 13.

a) Use information on the map to help you label the satellite image to show the Gulf of Mexico, Caribbean Sea and Tikal National Park.

b) Add dashed lines to the image to show how wildlife corridors link to the Tikal National Park.

c) How might the wildlife corridors help increase biodiversity in the Tikal National Park?

Revised

Stretch and challenge

The creation of wildlife corridors in Central America is an example of an 'international conservation project'. Explain why.

Revised

Other attempts at sustainable management of rainforests

Revised

National Parks

These are set up to protect the forests and ensure that damaging activities are stopped.

Medical Reserves

Pharmaceutical companies buy large areas of forest to prevent them being destroyed. They use the forest to find cures for diseases that affect people.

Biosphere Reserves

These are similar to National Parks. Increasing amount of human activity is permitted the further away you travel from a central area of unspoiled rainforest. For example:

rainforest → hunting and collecting allowed → farming and wood gathering allowed → settlement allowed → unprotected forest

Ecotourism

In Costa Rica, the 'Certificate of Sustainable Tourism' is awarded to tourist developments that help protect the environment, provide employment for local people, use local resources and sustain local culture. A major success has been 'Cayagu', a tourism group that actively promotes sustainable practices in its six developments including resorts, tourist lodges and hotels.

Why is sustainable management of the rainforest needed?

The world's tropical rainforests are very important to life both inside and outside the rainforest area. At a local scale they provide a habitat for the most diverse ecosystem on the planet, while at a wider scale they are the 'lungs of the planet'. Large forested areas absorb carbon dioxide and help prevent its accumulation in the atmosphere. They are also capable of affecting rainfall patterns both locally and over a wider area. Thus sustainable management of the remaining areas of the tropical rainforest ecosystem is essential.

Exam tips

You need to know the specific details of one ecosystem and the way people use it as a case study for your first examination. Create a case study revision card based on the idea on page 27, either for the Central American rainforest ecosystem or another that you have studied. Include the following information and don't forget that *specific detail* is required:

- a sketch map to show its extent and key physical and human detail
- a description of the natural processes that occur within the ecosystem
- the ways in which the ecosystem provides benefits for people
- the impact of people's activities on the ecosystem
- the attempts to sustainably manage the ecosystem.

The issue of desertification

One of the case studies you need to know for the examination is about desertification. **Desertification** is the process by which previously fertile land is changed into land too barren to farm. Overall, the death of vegetation in such areas will expose the soil to greater erosion by both wind and rain. The soil will also lose its supply of nutrients. Without soil these areas will not support farming. However, reasons for dying vegetation are a complex mix of natural and human factors.

This section will help build up your case study notes by giving you non-specific information about desertification, so that you can then add specific details about the area you have studied.

Knowing the basics

1. Using your knowledge, an atlas or the internet to help you, label the following deserts on Figure 14: Sahara, Thar, Mongolian, Great Australian, Atacama, Arabian.
2. Highlight the area affected by desertification that you are studying. Label the map with its name.

Revised

Regions at risk of desertification

Revised

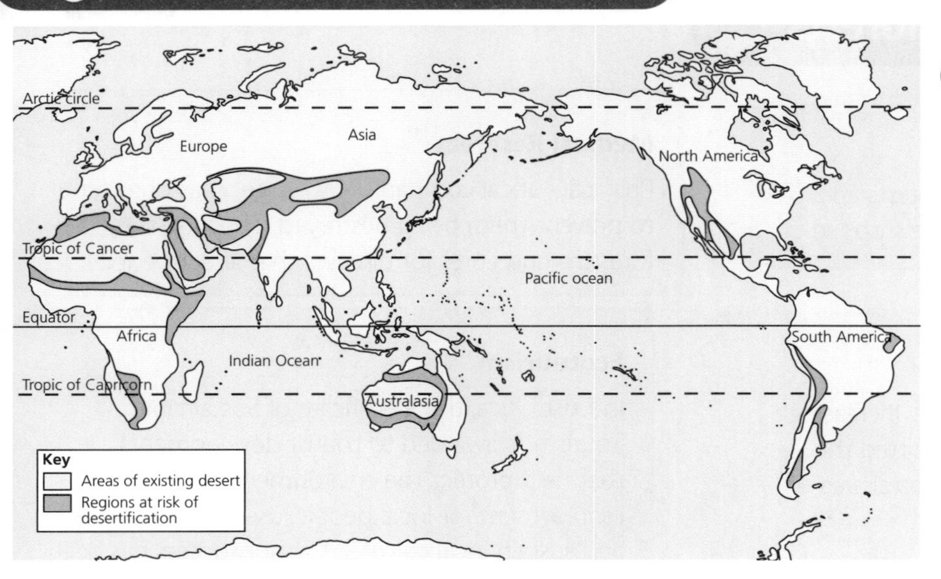

Figure 14 Regions at risk of desertification

Stretch and challenge

What is the relationship between current deserts and areas at risk of desertification? Think about ideas more specific than, for example, the deserts are 'near' or 'close' to them.

Revised

Water table – the level below which the ground is saturated with water

Knowing the basics

Why are many deserts found in western coastal areas?

Revised

Natural causes of desertification

Revised

Non-specific information	Detail specific to my study area
High pressure systems With changing weather patterns, some areas on the edges of deserts will be influenced more by high pressure systems. This will mean less annual rainfall and an increased likelihood of drought years in which plant life will decline. Many areas have a distinct dry season and this season becomes longer.	

Evapo-transpiration Increased influence of high pressure systems will result in more clear skies and an increase in high air temperatures. This will increase both evaporation from water surfaces and transpiration from the leaves of plants, resulting in lower **water tables** and less available water.	

Human causes of desertification

Revised

Non-specific information	Detail specific to my study area
Over-grazing Too many goats or cattle grazing an area of land will remove grass more quickly than it can regrow. The hooves also compress the soil, stopping air and water entering it and making it less fertile. This is especially a problem in societies where the number of cattle a family owns is a measure of wealth.	
Poor land management This includes overgrazing. It also includes taking too much water from wells for irrigation, causing the water table to continually lower until it is below the level of the wells. In economically poor areas there is also an inability to fertilise crop lands and to use water-efficient irrigation methods. At the other extreme, some transnational companies (TNCs) are involved in commercial farming that places too much strain on the soil.	
Plants for firewood For many people, there are few alternatives for cooking and heating. In many societies dried cow dung is used. This is not always available in sufficient amounts, so trees and bushes are destroyed for use in heating. Villagers travel increasing distances in search of the wood. Much wood is also used for building purposes. The soil becomes unprotected and erosion takes place.	

Managing desertification

Revised

Attempts to manage desertification are many and varied, and take place at scales ranging from local to global.

- Low lines of stones (bunds)
- Reforestation schemes
- NGO local community farm aid schemes
- International climate agreements
- Creation of shelter belts
- National water management schemes

Knowing the basics

1. The text boxes on the left show ways in which attempts are being made to manage desertification. Write a paragraph to describe each way that applies to your case study.
2. Do the same for any other strategy used in your case study that is not shown above.

Revised

Stretch and challenge

How likely is it that each strategy you have described will be successful?

Revised

How does the hydrological (water) cycle operate?

The water cycle

Revised

The water cycle comprises all the **stores** and **flows** of water. Water may be stored, or flow in or between, its three states: liquid, solid and gas.

Figure 15 shows part of a river's catchment area. This is the area from which it receives all of its water. Catchment areas are separated from each other along watersheds, which are ridges of higher land.

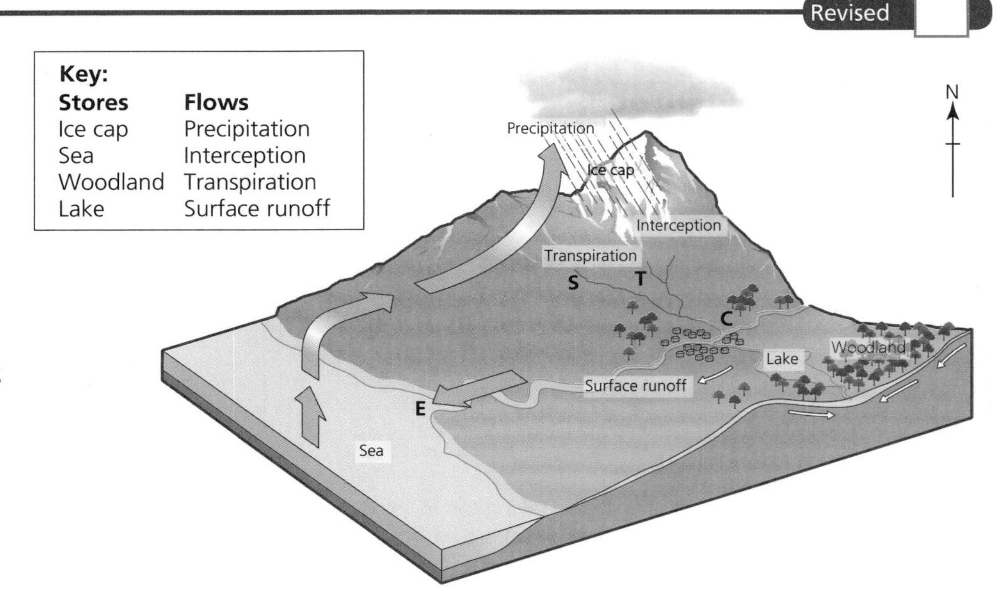

Figure 15 The hydrological (water) cycle in action

Water cycle variables

Revised

How water passes through the cycle depends on a number of variables. Woodland, for example, will increase evaporation and transpiration but will reduce surface runoff (overland flow).

As with vegetation, the nature of the rock in the catchment area will have a major influence. Impermeable rock will not allow water to pass through it and so will reduce the amounts and speed of groundwater flow. On the other hand, permeable rocks will allow the rapid passage of large quantities of water through them. All of these variables interact to determine how quickly and in what quantities the water that falls on a catchment area will enter the rivers and, therefore, the likelihood of **flooding**.

Stores (water cycle) – where water remains in one place or state

Flows (water cycle) – the transfer of water between stores, in the same state or involving a change of state

Flooding – when water covers or submerges a place or area that is normally dry

Human interference

Revised

Most of the land area of the UK has been directly affected by the activities of people. For example, urban areas are largely covered by buildings and other impermeable structures. Rural areas are dominated by vegetation, although much of this has been planted by people or is managed by them.

The nature of the surface has a major effect on how water falling onto a drainage basin returns to the rivers and, eventually, the sea.

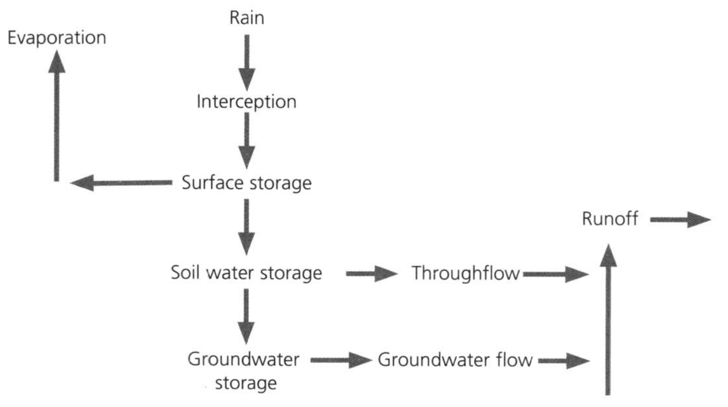

Figure 16 Rural discharge

Most water that falls on rural areas does so onto fields, moorland and woodland. Many of these surfaces are permeable so much water infiltrates the ground. The situation is quite different in urban areas. Here, lots of water falls onto impermeable concrete, tarmac surfaces, or the roofs of buildings. However, drains carry it quickly to the rivers. Increased urbanisation therefore increases the chances of flooding.

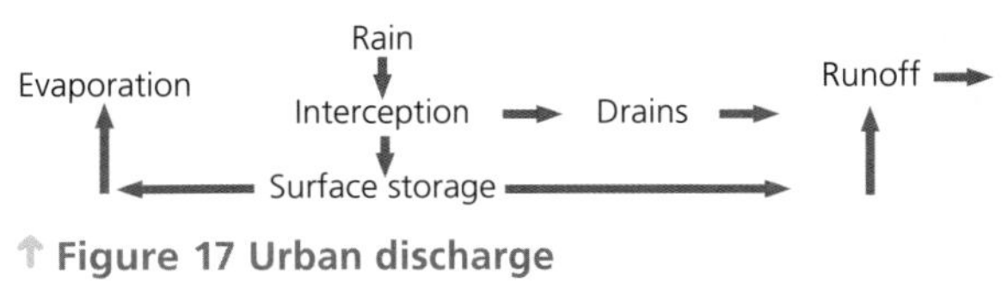

Figure 17 Urban discharge

Knowing the basics

Revised

Attempt these two matching exercises to revise some important terms.

1 Each of the letters on Figure 15 shows a feature of a drainage basin. Link each with its description.

A Confluence (C)	**1** where a stream begins
B Tributary (T)	**2** where two rivers join
C Estuary (E)	**3** a small river feeding into a larger one
D Source (S)	**4** where a river enters the sea

2 This is a more complex exercise. It asks you to link the names of the main flows of the water cycle with their meanings.

A Precipitation	**1** water seeping into the ground
B Interception	**2** transfer of water from a liquid to gaseous state
C Evaporation	**3** movement of water through the soil
D Transpiration	**4** water falling out of the atmosphere
E Infiltration	**5** movement of water through rocks
F Through flow	**6** transfer of water vapour from vegetation
G Groundwater flow	**7** where falling water hits the earth's surface

Stretch and challenge

Compare the discharge in rural areas with that in urban areas.

Revised

Flooding: an ever-increasing problem

Revised

Throughout history, lands on either side of some rivers have flooded as a result of the natural processes that take place in catchment areas. These areas are usually avoided for building and are often used for pastoral farming or recreation purposes.

However, in the UK, for example, a greater number of houses are built on flood plains (land that has traditionally flooded). This, together with the effects of increased urbanisation and changed weather patterns as a result of climate change, all increase the flood risk of some places.

River flooding

Severe floods

Revised

Some parts of the world are more naturally prone to severe flooding than others, for example deltas like those of the Ganges in Bangladesh and the Mekong in Cambodia and Vietnam. Both of these deltas are affected by a 'cyclone season'. Look back to pages 44 and 45 for the effects of cyclones.

In these areas people live with flooding every year but it is only occasionally severe. One such year for the Mekong delta was 2011.

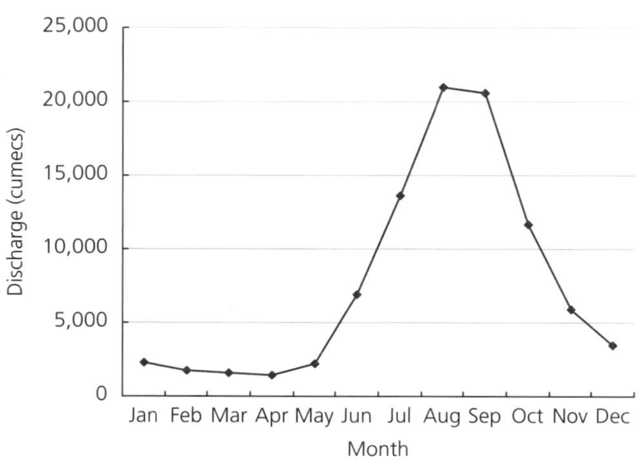

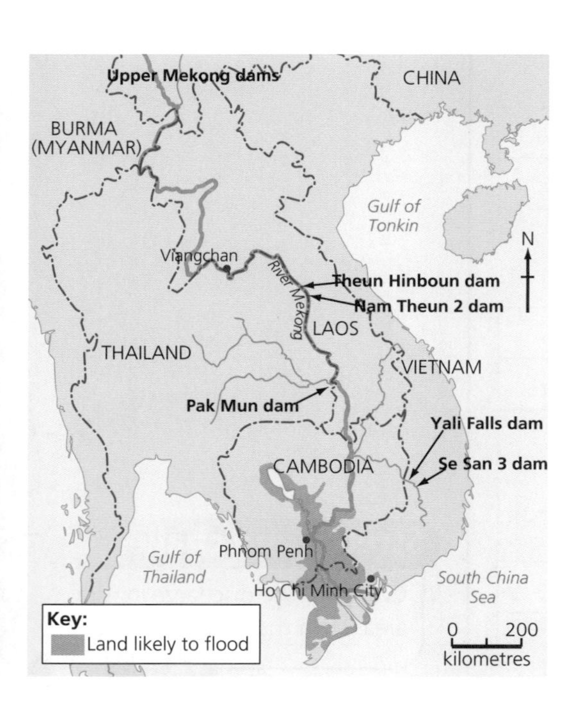

Figure 18 Flooding of the Mekong delta hydrograph

Direct from heavy rainfall 10%

River discharge from rain and snow melt 60%

Mekong Delta Flooding

High tidal flow from sea 30%

Exam practice

1 The bankfull stage along this stretch of the River Mekong is 9800 cubic metres per second (cumecs). Draw a line on the **hydrograph** to show this information. [1]

2 Complete the following passage using information from the hydrograph. [3]

Discharge of the River Mekong is low between January and ________. It then rises until August, reaching a peak flow of ____________ cumecs. From September to December discharge falls ____________ to 3500 cumecs.

3 Explain why the River Mekong floods. [4]

Answers online

Online

Exam tip

Question 2 in the Exam practice is typical of one that is used in the early parts of a Foundation Tier question. On the Higher Tier, the question would ask you to 'Describe the discharge of the River Mekong, using figures in your answer.'

Influences on river flooding

Revised

Snow melt

Snow accumulates in mountain areas during the cold winter period. As air temperatures rise in spring the snow melts rapidly to add large amounts of water to the discharge of rivers.

High water table

If the water table is high very little rainfall can infiltrate the soil. This means that much of the rain will quickly reach the river by surface runoff, increasing discharge rates.

Intense rainfall

A sustained period of heavy rainfall often brought by a severe storm like a hurricane or cyclone may cause flooding by falling directly onto the flooded area or onto the catchment upstream of it, causing higher discharge rates.

Deforestation

Trees intercept water and regulate its flow. Their removal causes greater surface runoff and also makes the soil susceptible to erosion. Soil washes into the river to collect on its bed, reducing its water-carrying capacity.

Urbanisation

Most water falling on urban areas finds its way quickly into drains through which it is channelled directly to the river, causing increased discharge and a rapid rise in water levels.

Effects of the 2011 Mekong delta flood

Revised

Type of loss	Loss and damage in the delta area
Human	89 people died
Built features	900 houses destroyed; 176,000 houses flooded; 1200 classrooms damaged
Farming	27,000 hectares of rice padis and 74,000 hectares of fruit trees ruined
Fisheries	7,300 hectares of fish farms and over 5,000 fish cages damaged. 5,600 tons of shrimp and fish lost.
Transportation	554 km of national roads and 316 km of rural roads destroyed; 9,000 bridges destroyed
Water control structures	4 million metres of dykes and 1.2 metres of irrigation canals destroyed

Knowing the basics

Revised

Produce a 'fact file' on the Mekong flood or any other flood you have studied.

Include information on:

- The location of the flood. Draw a sketch map to show it.
- The main causes of the flooding.
- The effects of the flooding on the lives of people.

Managing floods

Living with the annual flood

Revised

When it is too difficult or costly to control river flooding, the people living on flood plains must learn to live with it. This is the case in the Mekong delta area.

Figure 19 Flooding in the Mekong delta

International co-operation

Revised

The River Mekong is jointly managed by the countries through which it passes. They meet annually at the 'Mekong Flood Forum' to discuss how the river can be used in one part without causing problems for people along the banks of another part.

However, despite opposition from some other members of the forum, Thailand announced in late 2012 that it would support the construction of a large multi-purpose dam across the River Mekong along its border with Laos. This could result in:

Near the dam
- safer water supply
- cheap electricity
- economic growth
- greater water control
- cross-border sale of electricity

In the delta area
- a drop in water levels
- water release causing flash floods
- fewer fish
- less water control

Knowing the basics

Revised

Complete your 'fact file' on the Mekong delta by adding notes on river and flood management.

Stretch and challenge

Explain the importance of international co-operation in the management of rivers.

Revised

Preventing floods

Revised

It is important to know the difference between preventing floods and protecting people and environments from their effects. Look at the strategies around Figure 19 on page 64. All of these are **flood protection** methods which recognise that the river will flood annually.

Flood protection – actions taken to protect an area from the effects of flooding

Flood prevention is concerned with actually stopping areas of land flooding. Such strategies may either involve hard or soft engineering. **Hard engineering** involves building structures along the course of the river or altering the river's course. **Soft engineering** involves using knowledge of the catchment area and river features to stop the flooding.

Knowing the basics

Revised

1. In column 1 of the table below, name a place this method is used.
2. In column 2, state whether it is hard or soft engineering.
3. Add at least one positive and one negative to each strategy.
4. Which of the methods shown in the table is **not** a flood prevention method?
5. Look at Figure 20 on page 66. List ways in which the Trent flood plain is being used for purposes other than building. Which of the methods in the table is not a flood prevention method?

Stretch and challenge

1. Complete the final row in the table with one more flood prevention method
2. Which is the more sustainable, hard or soft engineering? Why do you think this?

Revised

Strategy	Hard/ soft	Positives	Negatives
Construct dams across river to hold back the natural flow of water.		Directly controls water flow.	Floods existing farmland.
Afforestation in the upper catchment area. Trees store rainwater and slow its flow.		Provides a supply of wood.	Replaces high biodiversity habitats with low.
Build walls and levees along a river's banks to raise its bankfull level.		Allows building on the flood plain.	Flooding is severe if they break.
Deepen the river channel to increase the amount of water it can carry.			
Use flood plain land for purposes other than building.			

What would you have done?

Recently a stretch of the River Trent west of Nottingham received a new flood wall. There were two options for its route in the commuter village of Attenborough.

The views of the villagers were not all the same.

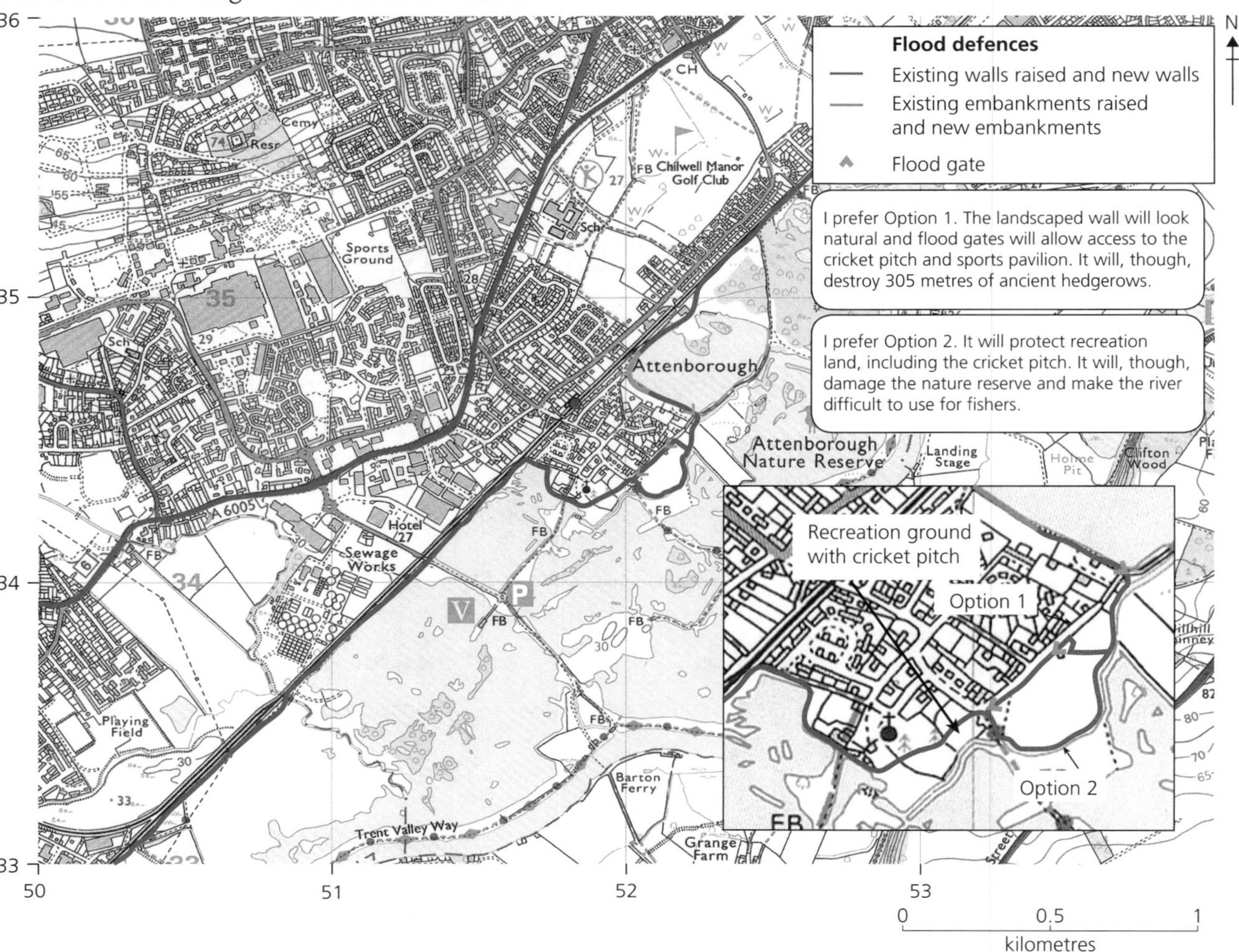

Figure 20 OS map extract showing the new flood wall on the River Trent

Exam practice

This is a mini problem solving exercise similar to the one that you will get at the end of Unit 2B, the problem solving paper. Do you agree with the chosen option? (Use evidence from this page and your own knowledge in your answer.

(This would be worth 11 marks on the Foundation Tier paper and 14 marks on the Higher Tier paper.)

Answers online

Online

What actually happened?

Figure 21 The cricket pitch before building began. November 2000: NE from GR. 520344.

Figure 22 The cricket pitch on completion of the flood wall. February 2013: NNW from GR. 521341

How are landforms produced?

Whether they are coastal or created by rivers, landforms are formed due to processes of erosion, transport and deposition operating in a similar way.

Agents of erosion

Revised

Figure 23 Agents of erosion

Differential erosion

Revised

Where rocks of different resistance are found together, erosion differs according to the resistances of the different rocks. The weaker rock recedes much more quickly than the stronger rock.

This also occurs when one rock type has lines or zones of weakness in it, as at Etretat on the north coast of France (see page 68). Here erosion has caused the following sequence to take place:

wave cut notch → cave → arch → stack

As the cliff retreats, it leaves behind an area of flat rock (known as a **wave cut platform**) at sea level.

Many areas of coast are composed of rock that has little resistance to erosion. The till (boulder clay) cliffs of the Holderness coast of Yorkshire (see page 70) are an example. Here, the average rate of erosion is about 2 metres a year. A different erosion process contributes to this very fast rate of coastal retreat. While waves are attacking the base of the cliff, rainfall soaks into the till from above. This bonds with the clay and increases its weight. Large amounts of the till slip down the cliff face under the influence of gravity. This is called **rotational slumping**.

Erosion in action

Revised

Knowing the basics

1 Label Figure 24 to show:
  - **a)** three vertical lines of weakness in the cliff
  - **b)** the sequence of erosion from notch to stack
  - **c)** the wave cut platform.

Stretch and challenge

Annotate Figure 24 to help explain how the agents of erosion operate on the headland.

Revised

Knowing the basics

Label and annotate Figure 25 to explain erosion of the cliff in the photograph.

Revised

↑ **Figure 24 Differential erosion at Etretat, N France**

↑ **Figure 25 Erosion of a till cliff, Holderness coast**

River erosion

Revised

Remember that the same processes of erosion operate along the length of a river. This is the same for transport and deposition. The relationship between these three processes and the features they produce depends upon the stage of the river from source to mouth. These stages are called 'courses'.

	Course of river		
Feature	**Upper**	**Middle**	**Lower**
Gradient	Steep	Gentle	Very gentle
Channel	Shallow and narrow	Wider and deeper	Wide and deep
Processes	Much friction and so rapid erosion. Attrition has not had much effect yet	Less friction and reduced erosion. Much transport	Little friction. Mainly transport and deposition
Discharge	Low	Increasing	Highest
Bedload	Large. Sub-angular stones and boulders	Smaller and more rounded	Small and rounded
Landforms	Interlocking spurs. Rapids and waterfalls	Gentle meanders	Wide meanders. Ox-bow lakes. Flood plain.

While the relationships in the table are true of many rivers, they are generalisations.

Look at Figure 26. It is a photograph of two of the world's most famous waterfalls, the Niagara Falls. The Niagara River flows from Lake Ontario to Lake Erie over a resistant horizontal band of limestone; the cap rock beneath this is a much less resistant layer of shale. The force of water undercuts the shale. Eventually the limestone breaks along cracks and collapses under the force of gravity. Collapsed limestone rests in front of the Niagara Falls. Erosion at the base of the Horseshoe Falls creates a plunge pool while a scree of collapsed limestone rests against the Niagara Falls. The falls slowly erode back to leave a gorge through which the river flows.

Knowing the basics

1 On Figure 26:

a) Add labels at the end of the arrows to show; limestone cap rock; less resistant shale, collapsed limestone blocks; Niagara gorge;

b) Add arrows and labels to show the locations of two areas of undercutting and a plunge pool.

c) Annotate your sketch to suggest how the area will change in the future.

Revised

↑ **Figure 26 The Niagara Falls**

Discharge – the volume of water flowing through a section of river at a given time. It is measured in cubic metres per second (cumecs).

Bedload – the material carried by a river being bounced or rolled along its bed

Transport and deposition

When the sea or a river has broken up rock, it then moves the rock fragments to another place. When the energy is too low to move the fragments, they are deposited. These processes are known as transport and deposition.

Transport

Water transports material in four different ways: **solution**, **suspension**, **saltation** and **traction**.

Solution – Rock material dissolves in the water. It may eventually become solid again if this water evaporates.

Suspension – Material 'floats' in the water as it moves.

Saltation – Material is bounced along by the moving water.

Traction – Material is rolled or dragged along. It does not leave the surface as it is carried.

River transport and deposition

Revised

Meanders: a balance between erosion, transport and deposition

A meander is a large bend in a river. The speed of water flow is greater on the outside of the bend. Here the banks are eroded to form a river cliff (1). Water speed is much slower on the inside of the bend. Here deposition takes place. The deposited material forms a low mud or sand bank; a slip-off slope (2). In times of high rainfall or snow melt the river reaches bankfull stage and overflows its banks. Material deposited on the valley floor creates the river's flood plain (3).

↑ **Figure 27 A meander on the River Severn**

Knowing the basics

Revised

1. Read the description of a meander above. Label Figure 27 to show features 1, 2 and 3.
2. Explain how a combination of erosion, transport and deposition helps to form the meander.

Stretch and challenge

Revised

How might this meander develop in future?

Coastal transport and deposition

Revised

As with erosion, the transport of material along the coast is similar to that along the courses of rivers. Solution, suspension, saltation and traction also operate along coasts.

Some material eroded from the coast is carried away from the shore and deposited at a distance. However, a large proportion of the material is transported along the coast by a process known as longshore drift. As it rests it produces a number of features of deposition. These include beaches and spits.

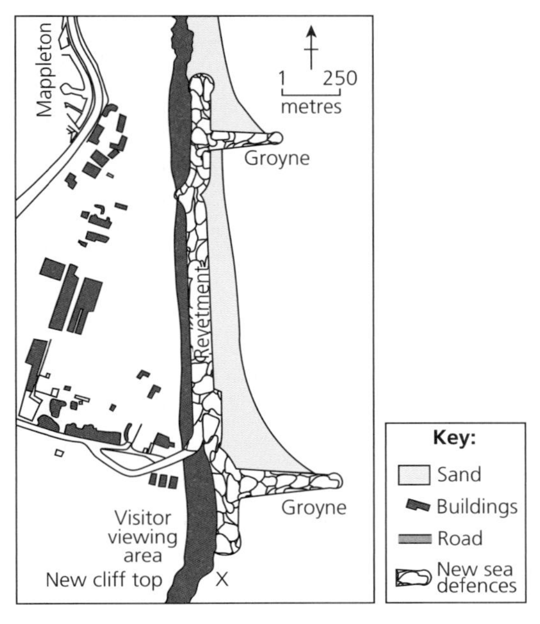

Figure 28 Beach processes and cliff protection at Mappleton

Longshore drift

Beach material is carried up the shore at the angle of prevailing wave movement. This is the **swash**, which is from the north at Mappleton. When the wave breaks, the water returns to the sea at right angles to the coast, pulled by gravity. This **backwash** carries the beach material with it, only to be picked up by another wave, and so the process continues. People build wooden, concrete or stone **groynes** at right angles to the coast to stop this movement of beach material.

Knowing the basics

Revised

1. Draw labelled arrows on Figure 28 to show the process of longshore drift.
2. Label the photograph to show how groynes help prevent longshore drift..

Stretch and challenge

1. How will a build-up of beach material help prevent cliff erosion?
2. Why has erosion increased at Point X since the creation of the cliff protection scheme?

Revised

Looking north from the south groyne

Can we prevent coastal erosion?

Coastal protection methods

Revised

As with flood risk, both hard and soft engineering methods are used to help defend coastlines from erosion. Each of the methods shown in Figure 29 is used along the Holderness coast and in other parts of the UK.

A Recurved sea wall

1 Large material ranging from pebbles to boulders in a steel cage. Break force of waves in advance of cliffs.

B Rip Rap

2 Loose boulders placed in front of cliff or sea wall to reduce wave energy. Also called rock armour.

4 Inserting drainage to cliffs and reducing their angle of slope to prevent rotational slumping.

C Revetments

3 Timber or steel structures placed in front of cliff to reduce wave energy.

5 Reinforced concrete structures with an outward curved upper part to deflect wave energy back to sea.

D Gabions

E Cliff stabilisation

Figure 29 Methods used to help defend coastlines from erosion

Knowing the basics

1. Match photographs (A to E) with their correct text boxes (1–5).
2. Groynes help build up beach material in front of cliffs to reduce the effect of waves. Suggest other advantages of halting the movement of beach material along the coast.

Revised

Stretch and challenge

All of the strategies above are hard engineering attempts to prevent coastal erosion. Soft engineering methods, like the importation of beach material (beach replenishment), are also used in some places. Consider the advantages and disadvantages of hard and soft engineering.

Revised

The nature of the problem

Areas of very weak, resistant rock like the Holderness coast have been subject to rapid erosion for years. Since records began, a large number of villages have disappeared into the North Sea. There is also a line of present-day settlements in danger of following them.

Some settlements are protected because they are considered too valuable to lose, while others are left to erode away. Governments, both national and local, must decide whether the costs of protecting settlements are greater than the advantages of protection. This is called **cost benefit analysis**. As with other issues, people will have different opinions based on their own particular sets of circumstances and the value positions they hold.

'Hold the line' or 'Managed retreat'?

'Holding the line' involves using strategies like those discussed above to prevent erosion. **'Managed retreat'** accepts that erosion cannot be stopped, especially in a future of (potentially) rising sea levels and increasing frequency of destructive storms.

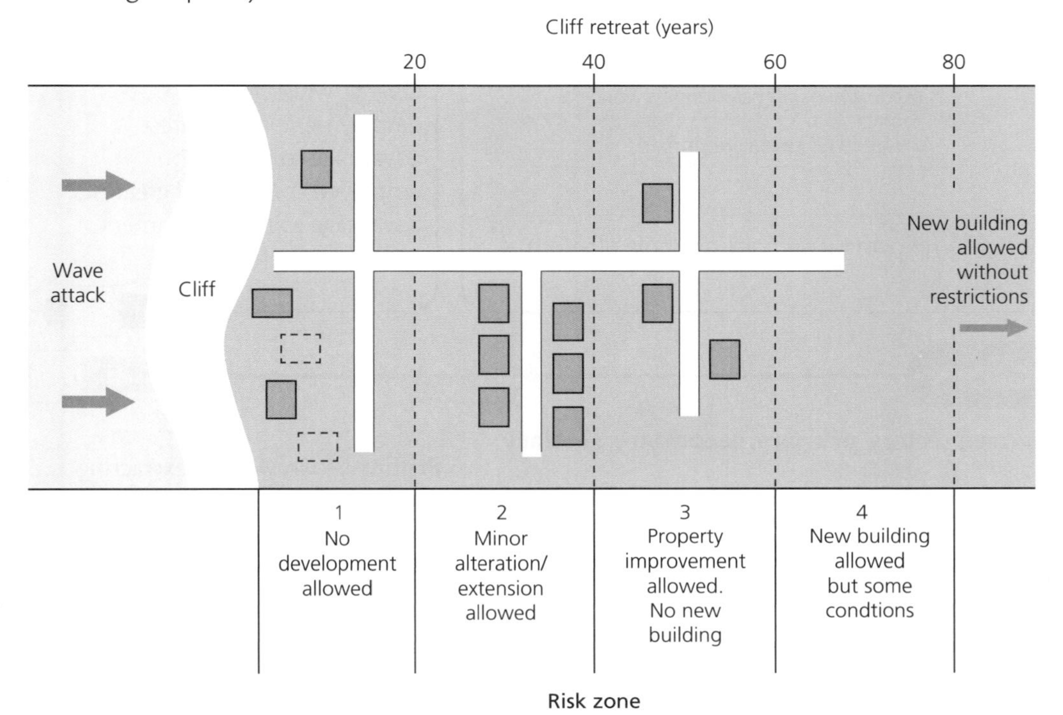

Figure 30 Managed retreat in action

Knowing the basics

1. Look at Figure 30. Are you in favour of 'hold the line' or 'managed retreat' along the Holderness coast? Why do you think this?
2. Who would agree with you and why would they agree?
3. Who would disagree with you and why would they disagree?

Revised

How and why do patterns of employment differ?

Formal and informal employment

Revised

Not all work is employment. You are working now, by preparing for an exam, but you are not being paid. That is the key – employment means receiving payment for work done. That's not all though. There are two types of employment: **formal** and **informal**.

Knowing the basics

Revised

1 Complete the table below to add one other feature of each type of employment.

Formal employment	Informal employment
Income tax taken out of pay	Income tax calculated at end of year
Have set times of work	Decide when and how long to work
Receive paid holidays	Any holidays taken are unpaid

2 Name one example of formal employment and one example of informal employment.

Stretch and challenge

Overall, which type of employment is of greatest advantage to the person employed? Which is of greatest advantage to the government? Explain your choices.

Revised

Employment sectors

Revised

People are employed in four main sectors: **primary**, **secondary**, **tertiary** and **quaternary**.

The proportion of these sectors varies according to where you are in the world and also the period of time the employment takes place.

Generally, in less economically developed places, more people will be employed in the primary industry than in the secondary and tertiary industries. The same pattern also applies the further back in time we go for any particular place. For example, the use of machinery in farming reduces the need for farm labour. Similarly, the recent trend of using computerised machinery in factories has resulted in less labour needed in secondary industries.

Primary – Growing or extracting raw materials.

Secondary – Manufacturing and processing goods.

Tertiary – Providing a service.

Quaternary – Providing information services.

Public sector – people employed by the national, regional or local government

Private sector – people who are self-employed or work for a larger company/organisation that is not controlled by the government

Knowing the basics

Revised

1 Look at the following ten jobs. Which jobs fit into each of the four sectors of employment?

farmer **steelworker** **fisher** **IT consultant**
miner **nurse** **car assembler** **secretary**
silicon chip maker **medical researcher**

2 Look back at the list of jobs above. List three jobs that are likely to be in the **public sector** and three likely to be in the **private sector**. Why is this task not as easy as it sounds?

Gender and age issues

Revised

Employment is not always fair. People are often treated differently according to whether they are male or female or according to their age.

From 6 April 2011, people reaching 65 no longer have to retire by default

Girls in poorer countries are less likely to go to school than boys

At the 2010 General Election 22 per cent of MPs elected were female

In some poorer countries, boys as young as 8 work long hours in unsafe factories

Figure 1 Unfair treatment?

Knowing the basics

Explain how the situation in each news headline above may affect the quality of life of individuals and groups of people.

Revised

Stretch and challenge

From the newspaper headlines above, select the two situations you think are most unfair. Suggest what might be done to make each situation fairer.

Revised

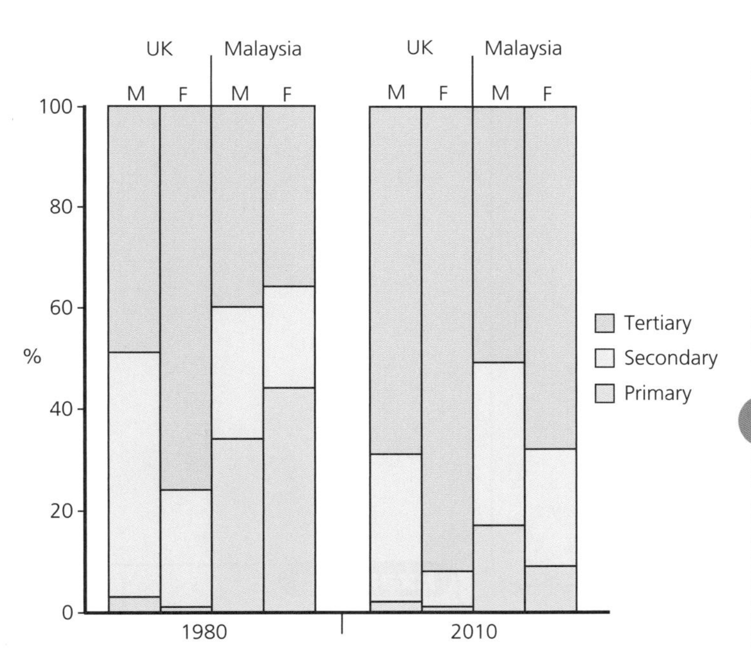

Figure 2 Shifts in employment by gender

Exam tip

You will be expected to respond to factors that influence the lives of people. In the exam, it is quite likely you will be asked to say how a particular factor could affect a person's quality of life. When answering a question like this, it is useful to try to put yourself in the position of the person or group affected ... and don't forget to use 'so what?' statements.

Knowing the basics

Explain how changes in technology may account for the following differences between the two years shown in Figure 2:

- a rise in females working in tertiary industry in both countries
- a much larger reduction of both females and males in primary industry in Malaysia.

Revised

Stretch and challenge

Revised

Add two more differences shown by the graphs and suggest why they exist.

Location, location, location

The importance of site and situation

Revised

Most of the world's industry is now controlled by **multinational companies** (MNCs). Sometimes, an MNC is located somewhere for simple reasons, for example an important mineral resource is found in a certain place and must be extracted from there. Most of the time, though, there is a complex decision-making process where all or some of the following factors need to be considered:

- availability of a suitable **site** for development
- the suitability of its **situation**
- transport links
- availability of a suitable workforce
- local or national government incentives.

Site – The land on which the industrial unit is to be built. Factors like size, flatness, and whether it has been built on before (brownfield) or is going to be built on for the first time (greenfield) may be considered important.

Situation – The land surrounding the site. Factors here include, for instance, availability of people for work, and whether there are nearby transport links, amenities and pleasant surroundings for the workforce.

Knowing the basics

Revised

The Cambridge Science Park was set up by Trinity College in 1970. Since then it has grown into the foremost research and development centre in the UK. Its flat greenfield site helped its development.

Complete the labelling and annotation of the sketch map in Figure 3 on page 77 to show why its situation has made it an important quaternary industry centre. Use information from the OS 1:50,000 map excerpt above it. Look at the inside front cover of this book to find the conventional signs (key) to the map.

Follow this sequence.

1. Label your map to show:
   - an A14 junction
   - Cambridge Airport
   - two Cambridge University colleges.
2. Shade and label:
   - a large built-up area
   - a country park
   - a greenfield site next to the Science Park.

Annotate your labels to show the importance of each to the company. Your annotations should all contain an elaborated 'so what?' element.

Stretch and challenge

Revised

Suggest how having an uninterruptible power source, being a telecommunications hub and having a 115-place nursery will help maintain the popularity of the Science Park.

Exam tip

It is possible to gain full marks for a case study by drawing, labelling and annotating a sketch map. Attempt to do this for either a secondary or tertiary industry you have studied.

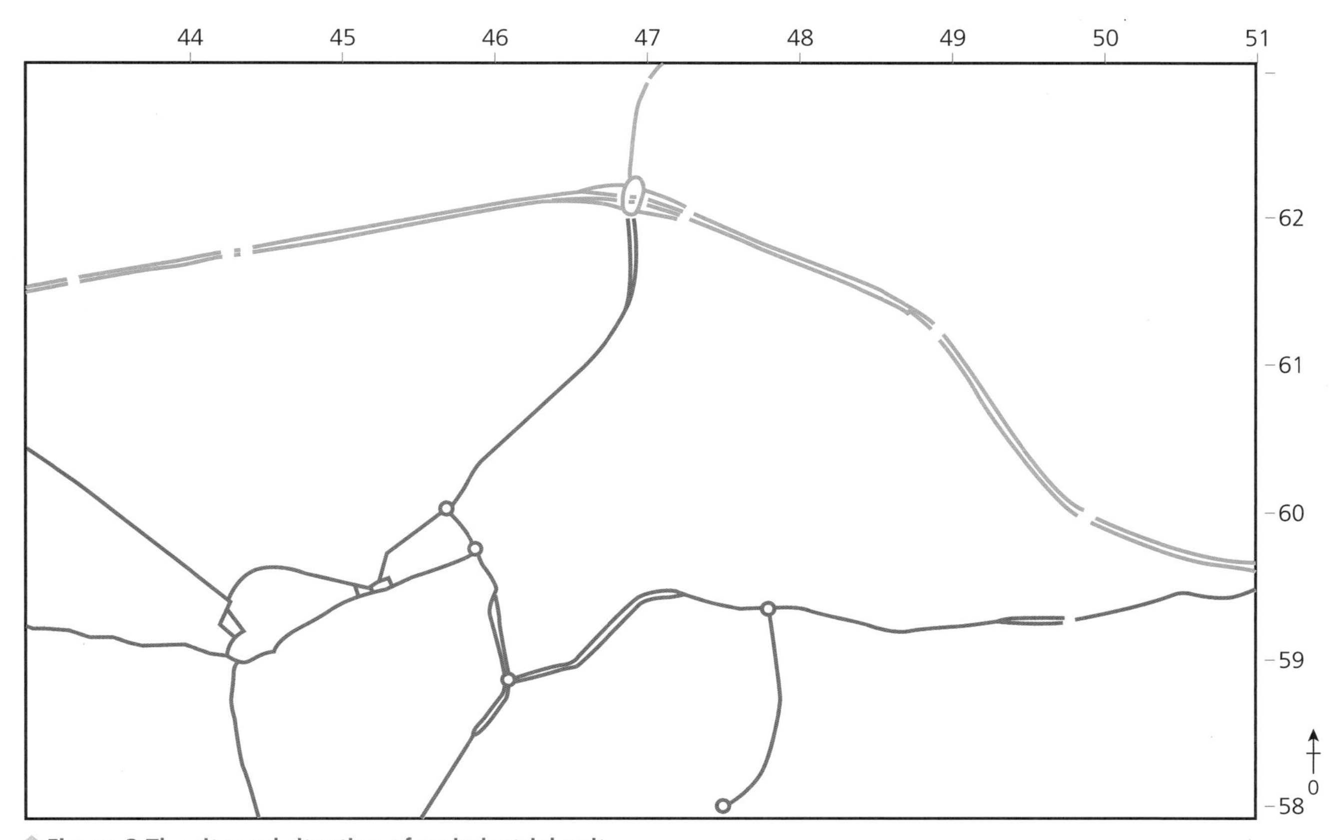

Figure 3 The site and situation of an industrial unit

Multinational companies (MNCs)

Rich companies and poor countries

Revised

The term '**global village**' was first used in the 1960s to describe how electronic communication, like the telephone, was making worldwide communications easier. It is now more broadly used to reflect the **interdependence** of countries through, for example, trade, travel, migrations, the internet and cultural links. The term '**globalisation**' refers to the way individual people, countries and industries are connected to each other on a global scale. MNCs (also called transnational companies or TNCs) are major influences on globalisation. These are large companies that have their head office in one city and other offices and factories around the world.

The profits made by the largest MNCs are many times greater than the Gross National Incomes (GNIs) of the world's poorest countries. For example, in 2011 the GNI of Sierra Leone was just $3 billion while the revenue of the WalMart retail group was almost $450 billion. Large MNCs are both rich and powerful.

Country	GNP per person (US$)
Democratic Republic of the Congo	348
Liberia	456
Zimbabwe	487
Burundi	615
Eritrea	735
Central Africa Republic	768
Niger	771
Sierra Leone	849
Malawi	860
Togo	899
United Kingdom	*38,669*

Ten poorest countries, 2011, and the UK for comparison

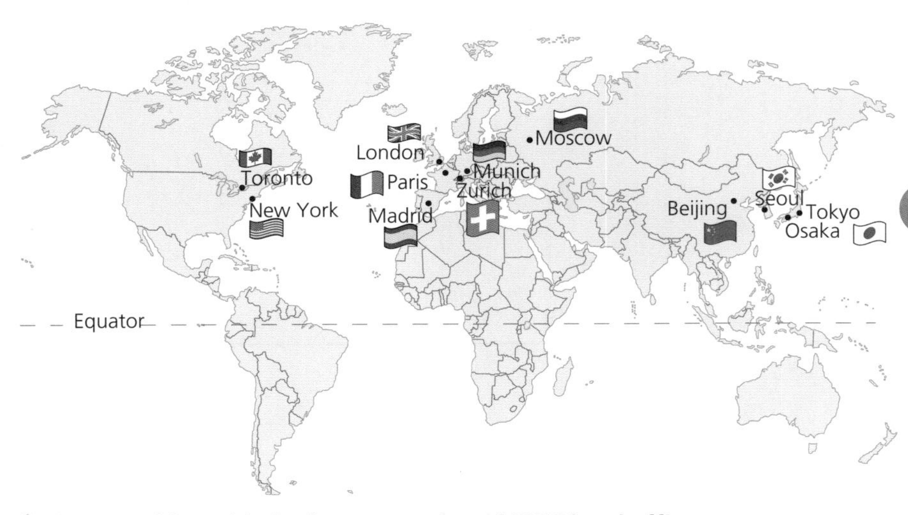

Figure 4 Cities with the largest number of MNC head offices

Exam practice

a) On Figure 4, shade the ten poorest countries.

b) Describe the distribution of the cities with the largest number of MNC head offices. Use figures in your answer.

c) Compare this with the distribution of the world's poorest countries.

d) Suggest reasons for the differences you have described.

Answers online

Online

Inertia or footloose?

Revised

An industry locates to an area because features of the area support its development. These features change with time, for example the source of energy used by the industry may run out. Some industries may have a large investment in the area, for example expensive immovable machinery, and will continue to operate there despite other features disappearing. This is known as **industrial inertia**. Many old heavy industries like steelmaking suffer from industrial inertia.

At the other extreme, **footloose industries** have fixed needs that don't change wherever they operate, and have capital machinery that is easy to move from place to place. This means that they may move around the world in search of the cheapest sources of their needs. Industries like computer chip or TV component manufacture are footloose.

Multinational companies: why locate there?

Revised

Figure 5 Location of BMW car factories

BMW is one of the world's leading car manufacturing companies. It has its head office in southern Germany and other car production and assembly factories at various locations around the world. The difference between a car production and a car assembly factory is that, in the former, the car is made from start to finish, whereas in the latter, the car is put together using parts made somewhere else.

Knowing the basics

Complete the following table for either BMW or for another MNC you have studied.

1 Add another location factor to the first column of the table and complete its explanation.
2 Use an atlas or the internet to help you name either a city or the country where the factory is located.

Revised

The **BRICM** countries (**Br**azil, **R**ussia, **I**ndia, **C**hina and **M**exico) are looked on as being both a source of cheap labour and production and, as they become richer, a target for increased sales. These and other rapidly industrialising countries are known as **newly industrialised countries (NICs)**.

Location factor	Explanation	Factory location
Goods made in EU member countries may be sold in any other member country without paying import duties	*so* the cost of the car reaching the buyer is lower than those made outside the EU so more profit can be made.	
UK local and national government offer incentives to companies setting up in the country	*so* ...	
Most MNCs wish to produce close to their headquarters to take advantage of a common language	*so* ...	
Manufacturers wish to make their goods close to major distant markets	*so* ...	
Labour costs are often high in the head office country but much lower in other parts of the world	*so* ...	
Another reason added by you ...	*so* ...	

Stretch and challenge

Revised

Suggest reasons for the locations of the BMW factories in Africa.

Multinational companies: good or bad?

The good?

Revised

As you have already seen, multinational companies bring a great deal of investment to an area. This results in direct employment – jobs for local people in factories – and often creates even more direct employment through jobs in associated industries. For example, a car assembly factory is likely to attract component manufacturers to the area.

The benefits do not end there. Investment of an MNC is likely to bring a positive **multiplier effect** to the area and sometimes to the country as a whole.

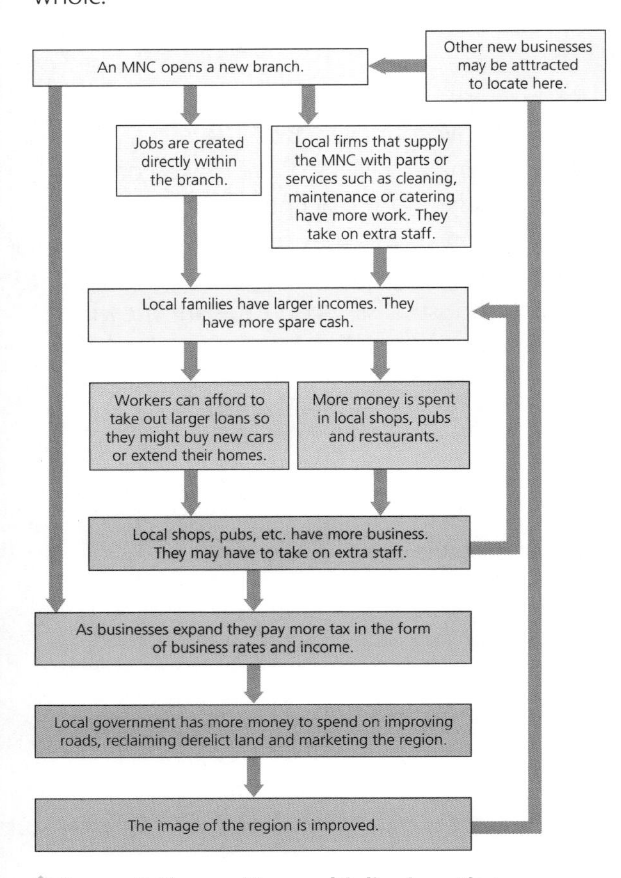

Figure 6 The positive multiplier in action

Knowing the basics

Figure 6 is a basic one. It looks at a process in general terms but does not include any specific detail about the effects of one MNC setting up a new factory or office.

Using information for an MNC you have studied, make a copy of Figure 6 but replace the generic statements with those specific to the MNC you have studied.

- Name the MNC and the place where the new factory or office was set up.
- Name new firms that have opened and the numbers of workers they employ.
- Give detailed explanations of local services that have benefited and what the benefits were.
- Describe the changes that have been made to the local area.
- Explain why these may attract more business to the area.

Revised

The bad?

Revised

Some people argue that multinational companies are not as good for an area as previously may have been suggested. The long-term records of some MNCs are not especially beneficial. Sometimes it is possible that the initial positive multiplier may, over time, turn into a negative.

More job losses in the South Wales electronics industry

Workers were told on Thursday that the LG Philips factory in Newport, South Wales, would close in August with a loss of 870 jobs. It produces components for computer monitors and colour picture tubes for televisions.

The large MNC blamed business and economic conditions for the decision. In recent years there had already been job losses in South Wales at Sony, Panasonic and Hitachi TV component plants. There are dangers when an area relies heavily on footloose industries for its employment.

Adapted from Mail Online, 14 February 2013

Why might it all go wrong?

1 Interfere in democratic elections

2 Answer only to their shareholders

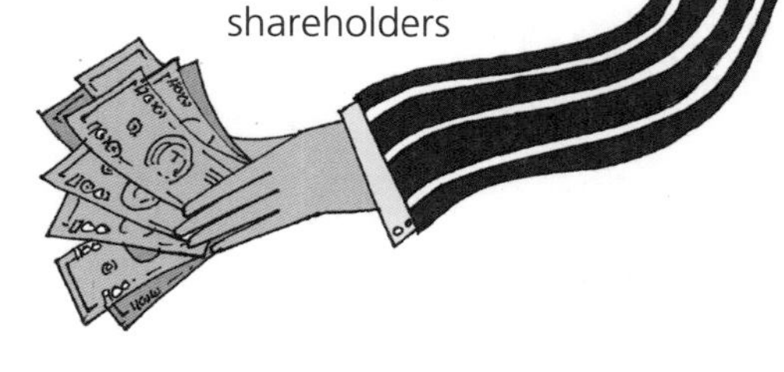

3 Have a poor human rights record

4 Cause much environmental damage

↑ **Figure 7 Some possible negative effects of MNCs**

A sustainable future?

Revised

Some national governments, international NGOs and, not least, consumers have pressured MNCs to improve their relationships with operating countries. As a result, there are now principles which state that there needs to be a more sustainable future for all parties involved.

Worldwide Responsible Accredited Production (WRAP) is an independent US-based organisation which certifies producers in the garment industry – an industry notorious for its performance in many poorer countries and one which supplies many well-known MNCs.

↓ **Figure 8 WRAP: Worldwide Responsible Accredited Production**

What WRAP demands

1. Compliance with local laws and workplace regulations.
2. Prohibition of forced labour.
3. Prohibition of child labour under fourteen years of age.
4. Prohibition of harassment or abuse and corporal punishment.
5. Compensation and benefits at least equal to the legal minimum of the country.
6. Working hours to comply with legal minimum, including at least one day off a week.
7. Prohibition of discrimination for any other reason than ability to do the job.
8. Health and safety at work and in factory-supplied housing.
9. Freedom of association and collective bargaining.
10. Environmentally conscious production.
11. Customs compliance to ensure no illegal shipments of goods.
12. Security to ensure no shipment of, for example, drugs or explosives.

Knowing the basics

Revised

It's your turn to decide. MNCs – good or bad?

1. List the positive and negative aspects of MNCs.
2. In your opinion, are they more of a good or a bad influence on the world?

Manufacturing industry: the environmental price

Different times, different places

Revised

The world's first industrial revolution took place in the United Kingdom in the late eighteenth and early nineteenth centuries. For the first time, people were able to mass-produce consumer goods and machinery to help with farming and other activities. People released from the land migrated to the cities that were growing around the coalfields, the main source of power.

Since then many other parts of the world have become industrialised. We call those countries that are industrialising newly industrialised countries (NICs). China is the largest of these.

Figure 9 Nineteenth-century pollution in Sheffield, UK

Local effects

Industrialisation leaves behind landscapes scarred from mineral and coal extraction and old factories often remain derelict in urban areas. Attempts to clear the land may not always be successful. In July 2009, for example, people gained compensation following attempts to clear up Corby steelworks, UK, following its closure in 1981. Women were exposed to toxic waste and, as a result of this, their children were born with defects.

Figure 10 Twenty-first century pollution in Beijing, China

Knowing the basics

Look at Figures 9 and 10.
Make a list of ways in which pollution like this can affect the lives of local people. Add an elaboration ('so what?' statement) to explain each effect.

Revised

The wider picture

Whether in the UK or China, the use of fossil fuels as a source of power creates problems that affect natural and built environments, as well as people. They are also likely to have not only local but international effects, such as damage caused by acid rain, and the enhanced greenhouse effect.

Figure 11 Acid rain

Statement	Position on diagram
Heavy particles drop onto the city, weathering stonework.	
Acid **throughflow** washes aluminium into rivers and lakes, clogging fish gills and disrupting the breeding of fish and frogs.	
Gases created by burning sulphur, nitrogen and carbon are created in power stations, factories and vehicles.	
Acid rain moves through the soil removing nutrients and adding aluminium. This kills trees and other plants.	
The main gases produced are soluble. They dissolve and travel great distances before falling as acid rain.	
Acid rain and fog damage the leaves of crops, making them stunted and giving low yields.	

Knowing the basics

Complete the table below to show the causes and environmental effects of acid rain. Place the correct number from Figure 11 in the right-hand column.

Revised

Stretch and challenge

Why is acid rain considered an international issue?

Revised

The enhanced greenhouse effect: a political hot potato

Revised

As schoolchildren were looking forward to their 2012 winter holidays and UK pensioners received their winter fuel allowances, the world's nations met in Doha, the capital of Qatar, to reach a new agreement on **climate change**. In a legally binding deal lasting until 2020 and extending the Kyoto Protocol (2005), rich nations agreed to compensate poorer countries for losses due to climate change.

However, the deal only involves 15 per cent of rich countries, and there appears little in it to effectively reduce greenhouse gases.

Below are some views that may help to explain why this is such a difficult issue.

Richer countries

'We have agreed an 80 per cent cut in greenhouse gas emissions by 2050 so why can't the poorer countries agree to cut by 50 per cent?'

'We will need to monitor the emissions of all countries to make sure they keep within the agreed limits.'

Poorer countries

'Our populations continue to grow and we need to develop our industries. The rich countries have already done this, so why shouldn't we?'

'We don't trust the richer countries. Monitoring emissions will give them the opportunity to gain valuable information about our industries.'

Up to date?

The news concerning the politics of climate change is constantly changing. Use the internet and radio, TV or newspaper reports to get the latest information. In the meantime, read on to explore its causes and effects.

What are the causes and effects of climate change?

Causes of climate change

Revised

Natural change

Climate change is certainly nothing new. In the past 2 million years Western Europe has experienced four major Ice Ages. Fossil evidence suggests that temperatures in southern England were tropical for some periods between these Ice Ages.

Human influence

Many scientists believe that the world is heating up more rapidly than has happened before because of people's activities, particularly over the past two or three centuries.

Effects of climate change

As the world heats up so patterns of air movements change. It is not simply a matter of everywhere getting hotter. Not only may people be affecting the world's climates but also world climate change may have major effects on people.

B. Deforestation leaves fewer trees to absorb CO_2. Burning trees also adds CO_2 to the atmosphere.

C. Burning fossil fuels in power stations, factories and vehicles adds CO_2 to the atmosphere.

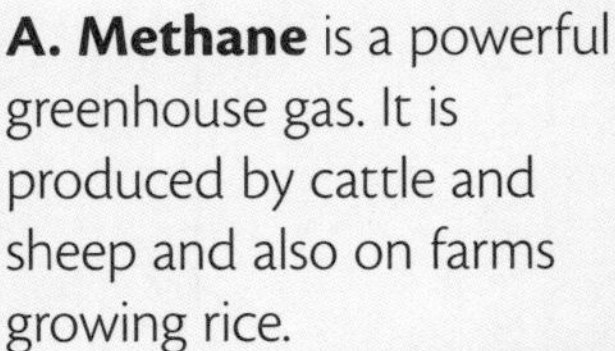

A. Methane is a powerful greenhouse gas. It is produced by cattle and sheep and also on farms growing rice.

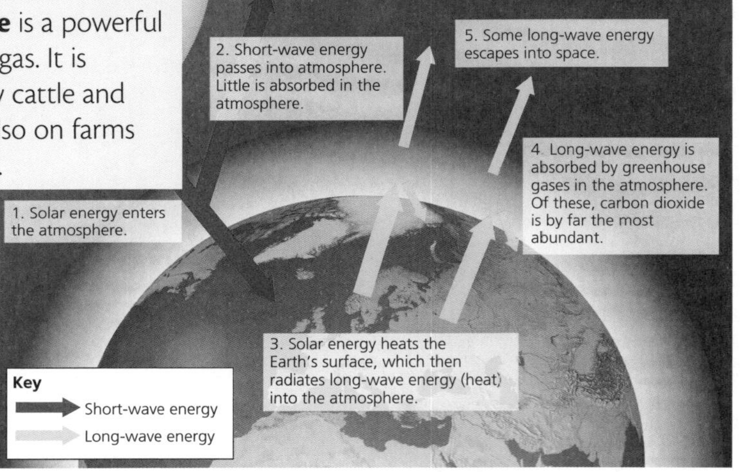

Figure 12 The greenhouse effect

Knowing the basics

1. Use information from Figure 12 to write a paragraph explaining how the greenhouse effect operates.
2. Why do some scientists believe that people's activities are affecting the greenhouse effect?

Revised

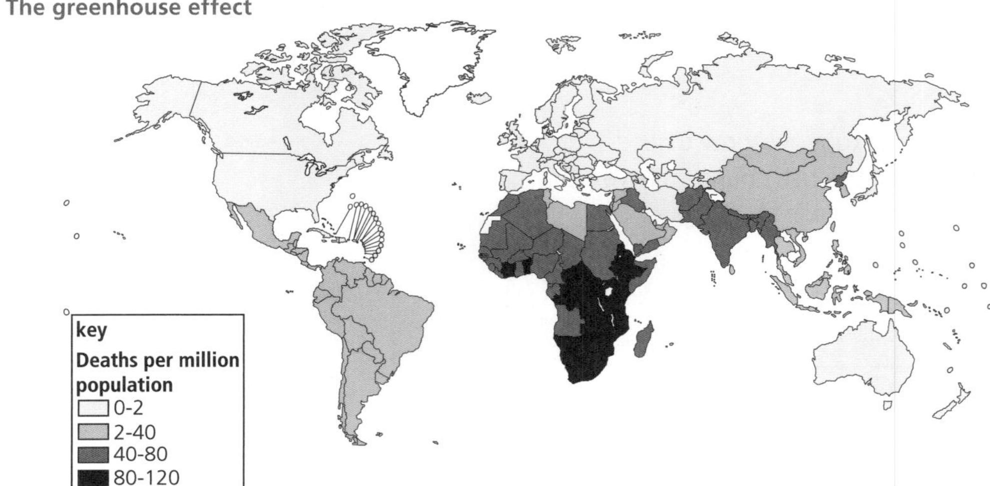

Figure 13 Estimated deaths per million population attributed to climate change in the year 2000. Source: WHO World Health Report 2002

DARA update 2010

The humanitarian group DARA estimated that, in 2010, around 5 million lives were lost worldwide because of air pollution and the effects of global warming. It estimated that the figure will rise to 6 million by 2030. According to DARA, the biggest impact was on China where 100 million people were affected overall and 1.5 million died, mainly from air pollution. They estimate that the loss of $72 billion dollars to the Chinese economy resulting from climate change alone in 2010 will have risen to $727 billion by 2030, because of major cuts in grain production and a great imbalance of water resources with increased drought and flooding.

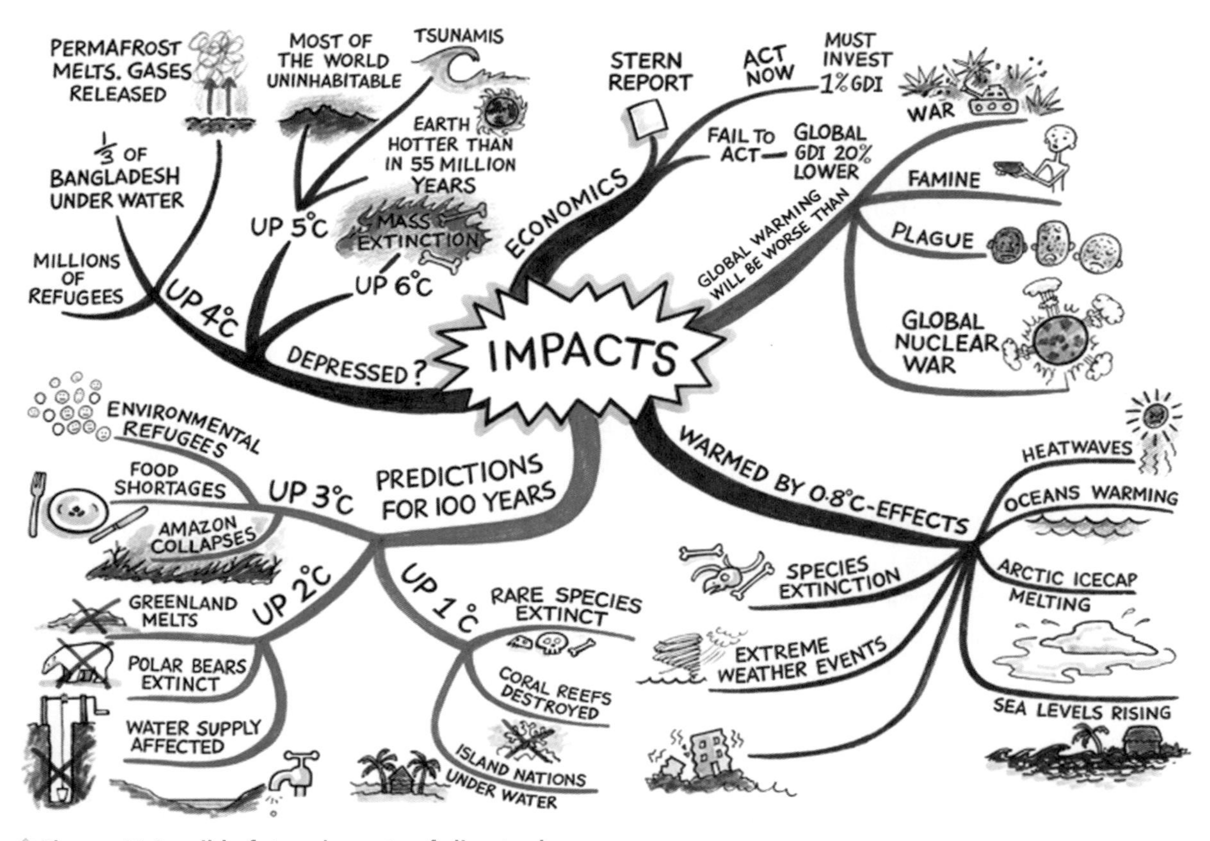

Figure 14 Possible future impacts of climate change

Exam practice

Although some questions found on the Higher and Foundation Tier papers are similar, there may be slight differences between them. Take these two questions:

Foundation Tier Explain why people need to act to reduce global warming. [5]

Higher Tier Explain why governments and individuals need to act to reduce global warming. [6]

Notice the greater demand of the question on the Higher Tier, where you must talk about two separate groups of people to attract the highest marks.

Attempt the question appropriate to you. Answer this question to the best of your ability. Information on pages 83–85 and your own notes should help.

Answers online

Online

Stretch and challenge

Use information on Figure 14 and your own knowledge to help label Figure 13 with different effects of climate change.

Revised

Sustainability: social, economic and environmental

Who decides what sustainability is?

Revised

People describe sustainability in different ways. The Brundtland definition on the right is commonly accepted but even this begs more questions than it answers. For example:

- What are the needs of the present? Are the 'needs' of a modern society, like private travel and the use of mobile phones, really necessary?
- What are the likely requirements of future generations? How can we tell?

Sustainable development is development that meets the needs of the present without compromising the ability of future generations to meet their own needs.

Source: Brundtland Report, 1987

As geographers we tend to divide sustainability into a number of separate areas:

Sustainable natural environments
For example, ensuring rainforests survive in the future.

Sustainable built environments
For example, ensuring that housing and service provision meet the changing demands of the people using them.

Sustainable economies
For example, ensuring that people of working age are employed and create enough income to meet the future needs of the unemployed and those too young or too old to work.

People rely on natural environments for their food, building materials and, either directly or indirectly, for their employment. It does seem, therefore, that both present and future social and economic needs will only be met if we first meet the needs of the environment.

What can we do?

Revised

Governments have agreed, in principle, to reduce greenhouse gas emissions but many of the targets look unlikely to be reached. How can people help?

Here are ten family strategies to help governments meet their targets:

- buy local produce
- use energy efficient electrical equipment
- buy from renewable energy electrical companies
- reduce heat loss from the house
- buy durable goods
- pack the fridge tightly
- walk or cycle more often
- eat less meat
- use public transport

Knowing the basics

1. Have you been counting? There are only nine strategies in the list. Add one more of your own.
2. Explain how each of the strategies will:
   - help the government reach its emission targets
   - benefit your family.

Revised

How *sustainable* are sustainable activities?

Revised

Reusing carrier bags when shopping

Boycotting furniture made from tropical hardwoods

Paying more tax for fuel

Taking holidays in the UK

Buying a low emission car

Reheating and using yesterday's leftover meal

Eating only local produce

↑ **Figure 15 Some sustainable options**

Knowing the basics

Revised

1 Look at the statements in Figure 15. Place them along a line, like the one below, according to how sustainable you feel each activity is.

Least sustainable → Most sustainable

2 Ask a friend or member of your family to do the same. Compare and discuss any differences.

Stretch and challenge

Revised

Add two other strategies to your line:

- one that is commonly used that you think is less sustainable than any of the strategies in Figure 15.
- one that is commonly used that you think is more sustainable than any of the strategies in Figure 15.

How might development be measured?

Economic indicators of development

Revised

Wealth is most often used to measure development. It is usually shown as the gross domestic income (GDI). This is the value of all final goods and services made within the boundaries of a country in a year. Sometimes this amount is divided by the country's population to give the GDI per person or 'per capita'.

Let's take an extreme example. A country of just four people earns \$40,000 in a year. The GDI of that country would be \$40,000 but its GDI per person would be \$40,000 ÷ 4 = \$10,000.

Gross National Income (GNI) is almost the same as GDI but is worked out slightly differently.

The Brandt Line, with which you may be familiar, is another way to measure development. It was drawn in 1980 to divide the world into rich and poor countries. However, much has changed since it was first drawn and nowadays the Brandt Line is rather out of date.

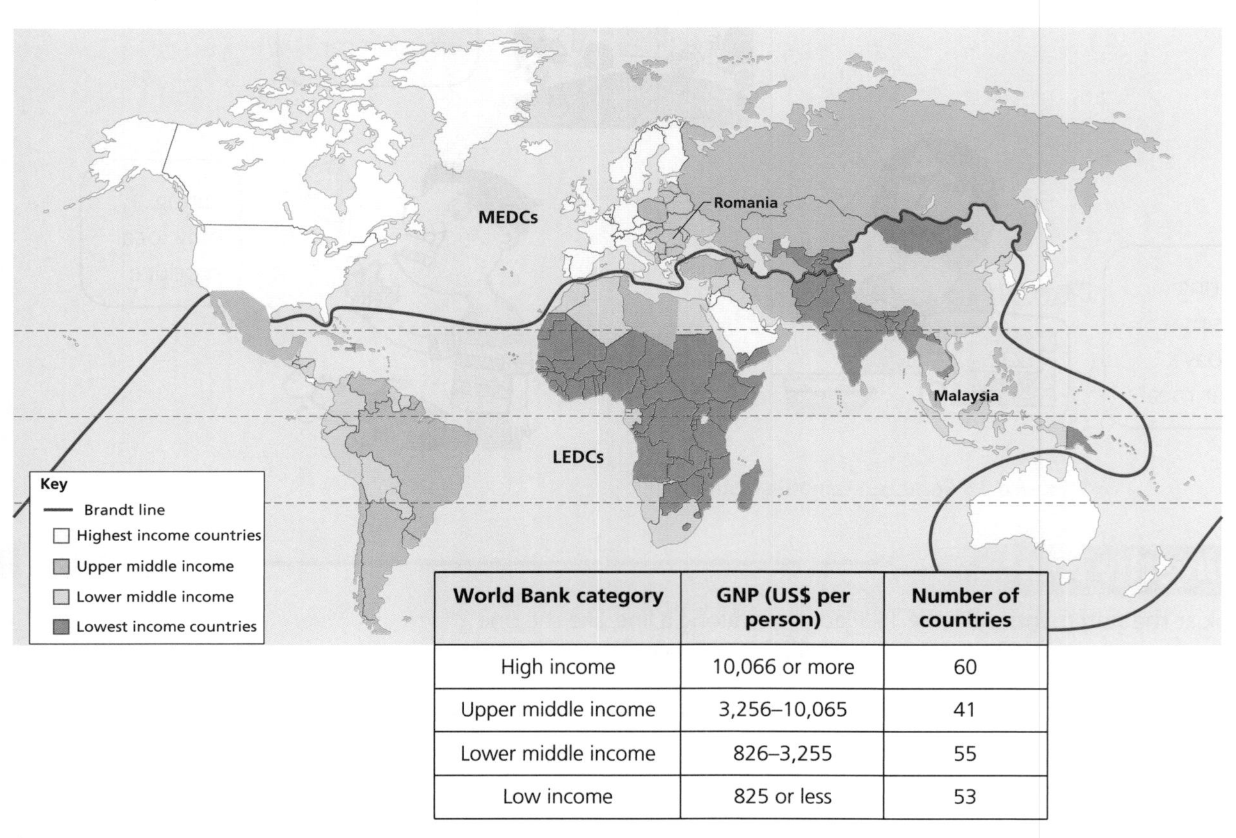

World Bank category	GNP (US\$ per person)	Number of countries
High income	10,066 or more	60
Upper middle income	3,256–10,065	41
Lower middle income	826–3,255	55
Low income	825 or less	53

↑ **Figure 16 The Brandt Line 30 years on**

Knowing the basics

1. Which two continents have the greatest proportion of countries with a very low income?
2. To what extent is the Brandt Line now a useful means of dividing rich and poor countries?

Revised

Stretch and challenge

One purpose of drawing the Brandt Line was so that the aid provided by rich countries could be targeted more effectively. On Figure 16, draw a new version of the Brandt Line. Explain your choice of line.

Revised

Development may be defined and measured in many ways. One way is to look at the wealth of a country, as seen in the GDI example above. However, wealth is not shared equally so even a wealthy country can have many people living in poverty. A more useful measure of development is to examine such factors as the standard of education, healthcare or political freedom enjoyed by ordinary people.

The Brandt Line only intended to divide richer and poorer countries, yet the countries to the north fell into two further groups. The communist countries of Eastern Europe and Asia developed much more slowly than the capitalist countries of Western Europe and North America. Romania is a former communist country, for example, while the UK is classed as part of the capitalist group. When the communist bloc collapsed in the early 1990s, Romania was far behind the **economic development** of the UK.

Recently, many of the poorer countries to the south of the Brandt Line have begun to industrialise rapidly. These are known as NICs (see pages 78–79). Malaysia is one of these.

Indicator	UK	Romania	Malaysia
% access to sanitation	99	89	96
% literate over 15 years of age	99	98	93
% of roads that are paved	100	50*	80*
Life expectancy at birth	79	70	72

↑ **Figure 17a Some key development indicators 2010 (* = 2005)**

Year	UK	Romania	Malaysia
2000	27	4	21
2005	65	17	49
2010	78	40	56

↑ **Figure 17b It's an increasingly smaller world: internet users as a % of total population**

Knowing the basics

Revised

1. Look at the information in Figure 17a. Decide which you think are the least and most important indicators of development shown. Why do you think this?
2. Which other indicators do *you* consider important? Why?
3. Google the 'Human Development Report' or 'UNdata country figures' to find evidence for the indicators *you* have chosen.
4. To what extent does the information you now have support the views *you* formed about the Brandt Line?

Social indicators of development

Revised

All the indicators shown in figure 17a are economic. They are measures concerning a country's ability or willingness to spend money to the benefit of its people. However, development can also be about ways in which different groups of people are treated within a country.

The United Nations Human Development Programme issues a Gender Development Index based on indicators such as male/female comparisons of, for example, years of education, representation in parliament, and employment type and pay.

All of these are measurable but it is also possible to use more subjective indicators like 'happiness'. The idea of Gross National Happiness (GNH) was used first in the 1970s by the King of Bhutan. His original idea has developed into nine main areas: psychological well-being; health; education; time use; cultural diversity and resilience; good governance; community vitality; ecological diversity and resilience; and living standards.

Stretch and challenge

Which of the nine areas of GNH do you consider most important to your life? Why?

Revised

Interdependence

Positive and negative effects of interdependence

Revised

Countries depend on each other in many ways. Different parts of the world are so interrelated that a major event in one part is likely to affect many other places. Some effects of interdependence are positive while others are negative.

1. Immigrants required

Ottawa, 31 October, 2012 – The Government of Canada will maintain record levels of immigration to support economic growth in 2013. 'Our Government's number one priority remains economic and job growth,' said Minister Jason Kenney. 'Newcomers bring their skills and talents, contribute to our economy and help renew our workforce so that Canada remains competitive on the world stage.'

Citizenship and Immigration Canada plans to admit a total of 240,000 to 265,000 new permanent residents in 2013, for the seventh straight year.

2. Food travels well

The vibrant, intensely colourful world of Indian food in Australia found an ever increasing fan base when Australians began to travel through India during the 1960s and 70s. Each region of India has its own style of cooking and distinct flavours. North is known for Tandoori and Korma dishes, South is famous for hot and spicy foods, the East specialises in chilli curries, the West uses coconut and seafood and the Central part of India is a blend of all.

3. Floods affect tourism industry

Gabarone, 16 February, 2013 – The recent floods experienced in Botswana's northern region have caused significant damage to certain lodges, resorts and camps, resulting in an unfortunate negative impact for some businesses in the tourism sector.Over 450 mm of rain had fallen, with particular reference to the Mashatu area, with the Limpopo River having recorded the highest water levels in history.

For some tourists, reaching the Limpopo valley airfield has been a challenge. The usual mode of transport has also become impossible due to the overflowing rivers.

4. The M6 Toll road

'The M6toll is the most exciting development in British transport history for many years. Since the M6toll opened in December 2003, it has grown from strength to strength and is recognised as the key strategic through route in the West Midlands; bypassing one of Europe's most congested motorways.

Midland Expressway Limited (MEL) is a private company with the government concession to design, build, operate and maintain the 27 miles of the M6toll until 205[illegible]. After this time the road will be handed back to the government. MEL is 100% owned by the *Macquarie Atlas Roads*.'

Macquarie Atlas Roads is a MNC with its head office in Australia.

5. Promoting sustainable cocoa trade in Ghana

Swiss Broadcasting Corporation, 12 December, 2011 – Ghana is the most important supplier of cocoa for Swiss chocolate makers. But the future of cocoa growing in the west African country is uncertain.

That's why a joint Swiss-Ghanaian project is attemping to rejuvenate the cocoa industry which is facing the problems of a shortage of young farmers and plantations full of old trees.

6. A new dam for Africa?

The proposed $80 billion Grand Inga dam in the Democratic Republic of the Congo will generate twice as much electricity as the world's current largest dam, the Three Gorges in China. It will jump-start industrial development on the continent. Funded by the World Bank and large MNCs, outside technology will be used to generate electricity to export as far away as South Africa, Nigeria and Egypt, and even Europe and Israel. On the down side, it is likely to bring little benefit to people in the DR of Congo and cause immense environmental damage.

↑ **Figure 18 Factors that drive interdependence**

Knowing the basics

Link the six situations in Figure 18 with each of the following aspects of interdependence: migration, trade, investment, culture, tourism and technology.

Revised

Stretch and challenge

Suggest which situations in Figure 18 are positive features of interdependence and which are negative. Are some both positive and negative?

Revised

The development gap

Revised

The **development gap** is the difference in development between the world's richest and poorest countries. It is most easily recognisable in the contrasting GNIs of the countries (look back at the table on page 78). An economically poor country is at a disadvantage when trading with richer parts of the world. This is especially noticeable when the country depends upon a single product for much of its income.

Overdependence on one form of foreign income

Ghana is a country in West Africa. It is a member of the Commonwealth of Nations, has strong links with Europe, and is a large producer of cocoa.

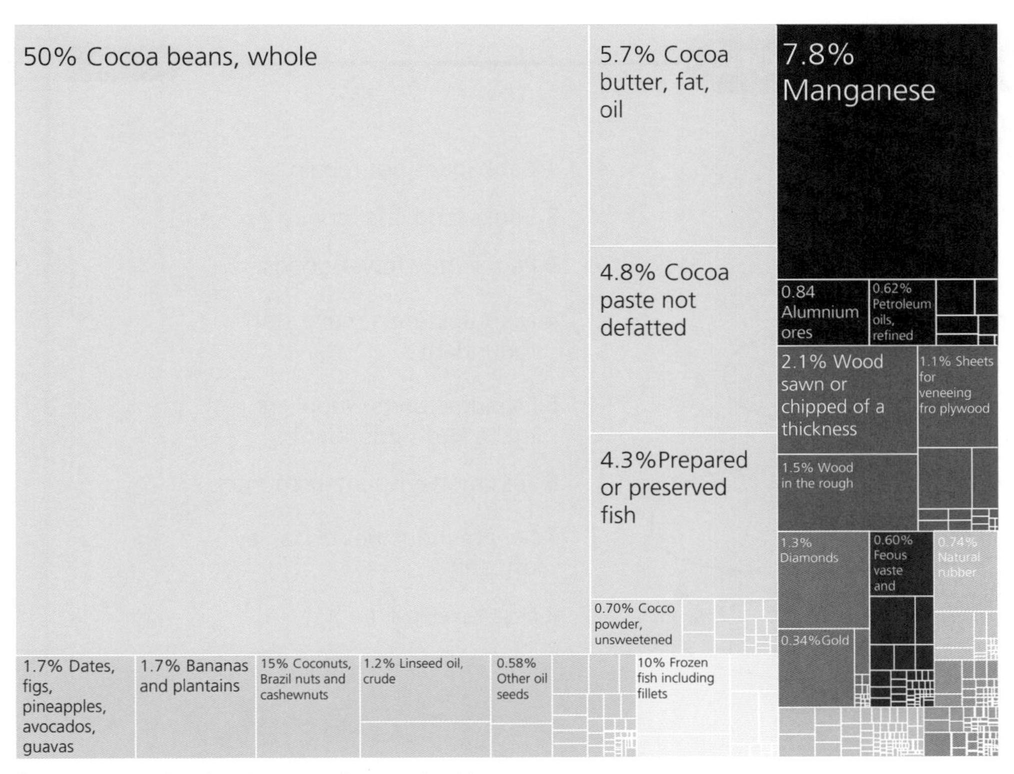

Figure 19 The dominance of cocoa in Ghana's exports

Knowing the basics

1. What proportion of Ghana's exports shown in Figure 19 is unprocessed cocoa?
2. What proportion of the exports is processed cocoa?

Revised

currency exchange rates

cocoa disease

bush fires

cocoa price changes

Cocoa export earnings

cocoa pests

drought

changed EU cocoa demands

Figure 20 Threats to Ghana's earnings from cocoa

Stretch and challenge

Why do countries prefer to process their raw materials before exporting them?

Revised

Knowing the basics

1. Which of the threats to cocoa earnings shown in Figure 20 are outside the control of the people of Ghana?
2. What evidence is there in Figures 19 and 20 to suggest that it is unwise for Ghana to rely as it does on exporting cocoa?

Revised

What are the benefits and disadvantages of an interdependent world?

In recent years the collapse of several banks in the USA quickly affected money markets in other countries, causing difficult times for borrowers. In the UK, as with other countries, this prevented people from buying houses, persuaded them not to buy new goods such as cars, and resulted in higher levels of unemployment. Such a period of time is known as a depression.

What happens in a depression?

Revised

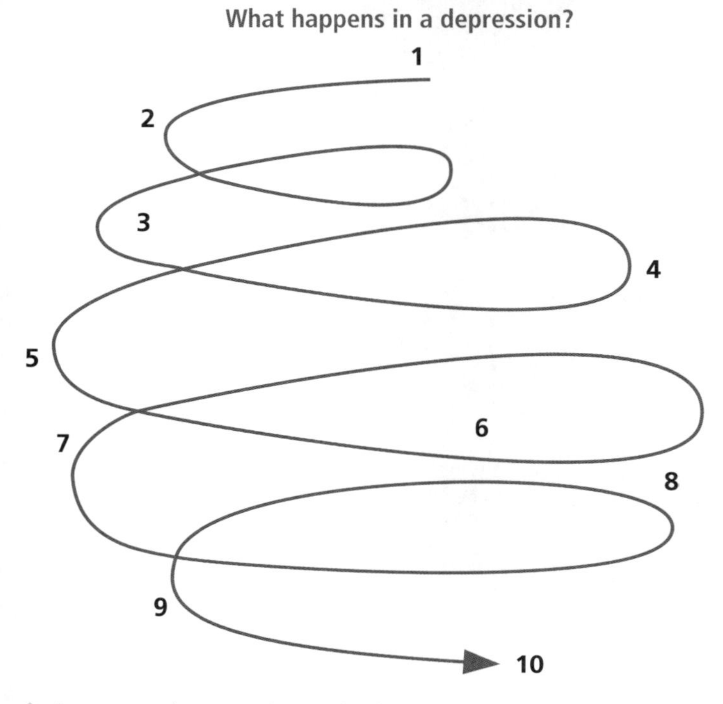

1 Bank loans not repaid

2 Banks lend less money

3 People buy fewer goods

4 Manufacturers make staff redundant

5 Manufacturers' suppliers make staff redundant

6 Less money spent in country

7 Service industries close, lay off staff

8 Less taxes paid

9 Less money for public services

10 Public services reduced, staff made redundant

↑ **Figure 21 The negative multiplier**

Exam tip

You will obviously be given the opportunity to use some of your case studies in the case study part of the three 'themed' questions (Units 1 and 2A of the exam). You may also be able to bring some of the information in as evidence for the decision you make on the problem solving paper (Unit 2B).

Wherever you use this information, you must give specific detail to gain the highest marks. For example, in this case:

- name actual places
- quote precise figures
- give detail of jobs lost.

To do this, you need to revise effectively and practise selecting the right information at the right time.

Knowing the basics

Use either a case study you have worked on in class or an example from your local area to outline the effects of the negative multiplier. Follow these stages:

1. Either above or on a separate copy of the negative multiplier diagram without the statements, add your own statements to include specific places, jobs and numbers of people affected.

Revised

Positive aspects of interdependence

Revised

The positive and negative aspects of interdependence above are at regional, national and international scales. How, though does it affect an individual – for example, you?

Knowing the basics

Revised

1 Complete the table below to show how your life is dependent on other parts of the world. Complete the blank rows to show other 'interdependent' influences on your life.

Article	Country
Article of clothing 1	
Article of clothing 2	
Hot drink	
Canned food	
Fresh fruit	
Electrical equipment	

2 How might your life change if goods and services were not exchanged between countries?

Stretch and challenge

It's not just you who is affected by this aspect of interdependence. From what you know already, suggest some advantages for the countries providing you with the goods and services you have listed.

Revised

It depends on your perspective

Revised

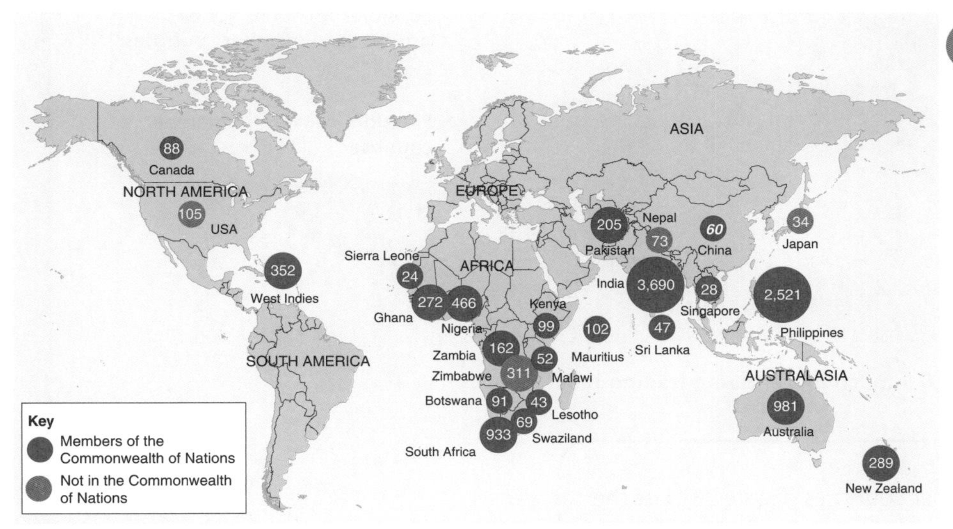

Figure 22 Newly qualified nurses in the NHS trained outside the UK, 2004–5

Knowing the basics

Most nurses have come from economically poor countries.'

'Most nurses have come from countries that are members of the Commonwealth of Nations.'

Why might these facts have influenced the movement of nurses to the UK?

Revised

Knowing the basics

Revised

On balance, is this migration good or bad? Think about effects on the country of origin and the country of destination. Who would agree with your view and who would disagree?

How does trade operate?

Trading agreements

Revised

As we have seen, much of the interdependence between countries is the result of trade. Traditionally, the UK has depended on the countries of the Commonwealth of Nations for its primary materials. In turn it has supplied these nations with manufactured goods and financial services. Nowadays, trade is a great deal more complex.

There are two basic types of trading agreement:

1. Free trade – the **import** and **export** of goods and services without any restrictions.
2. Restricted trade – this usually involves protecting a country's own industry by using some means of blocking imports from other countries, via **quotas**, **import duty** or sometimes the use of **subsidies**.

Imports – Goods that a country buys in from other countries.

Exports – Goods that a country sells to other countries.

Quota – A limit placed on the quantity of goods that a country may export to another.

Import duty – A tax or tariff that must be paid by a company when exporting its goods to a country.

Subsidy – A payment a country makes to its own producers to make them more competitive against imported goods.

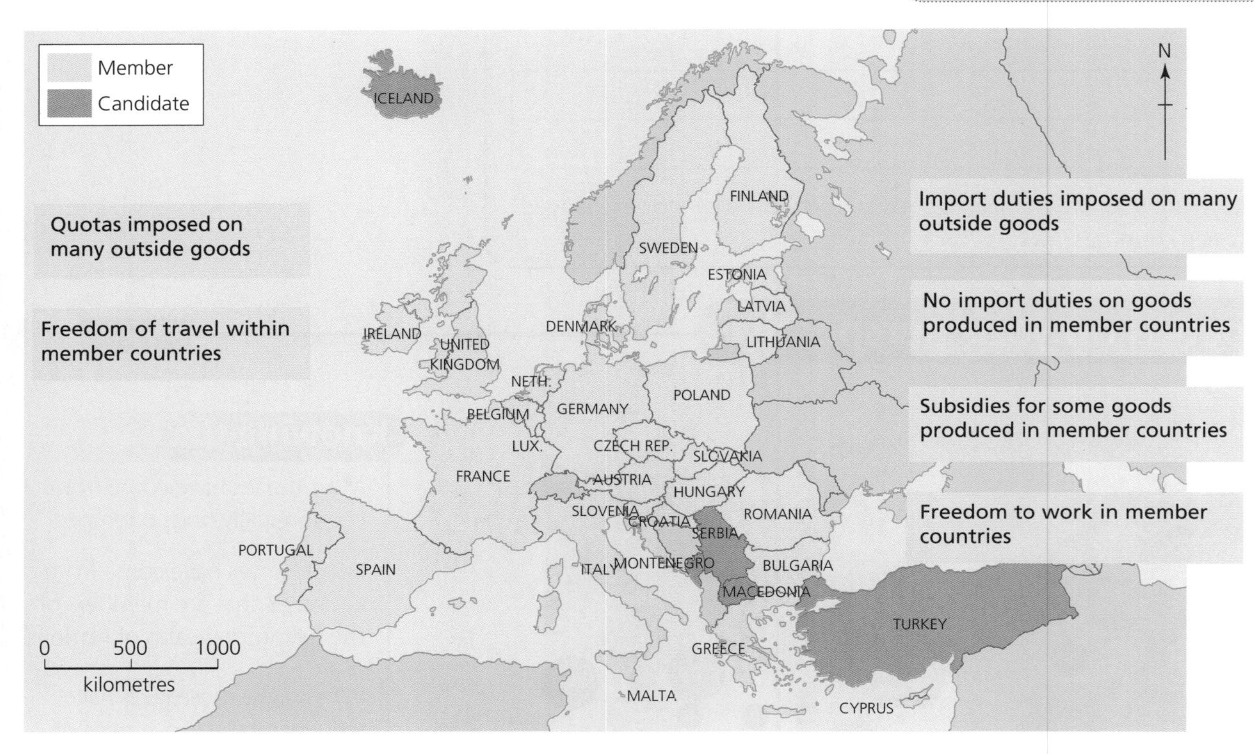

Figure 23 The European Union (EU) – much more than a **trading bloc**

Knowing the basics

1. Look at the map of EU member countries (Figure 23). List the trade advantages of being a member country of the EU. For each factor you have listed, write a sentence to explain the advantage for a firm manufacturing in the UK.
2. Some advantages apply to individuals living in an EU country. Use these advantages to explain how living in another EU country could directly benefit you in the future.

Revised

Stretch and challenge

Many of the EU member states also have a common currency, the Euro. What are the advantages and disadvantages of this?

Revised

Free trade: the law of supply and demand

Revised

On page 91, you explored some of the dangers of one country, Ghana, relying heavily on one export, cocoa. Cocoa is made into chocolate. People in the western world eat a lot of chocolate. Most of Ghana's cocoa is exported to the USA and to EU countries, so western chocolate companies buy the cocoa for the lowest price they can pay. The average income of a cocoa farmer is only about £160 a year.

A surplus of cocoa means there is more cocoa than is required, so the price paid to Ghana's farmers goes down. When there is a cocoa deficit there isn't enough cocoa to meet the demand, so the price goes up.

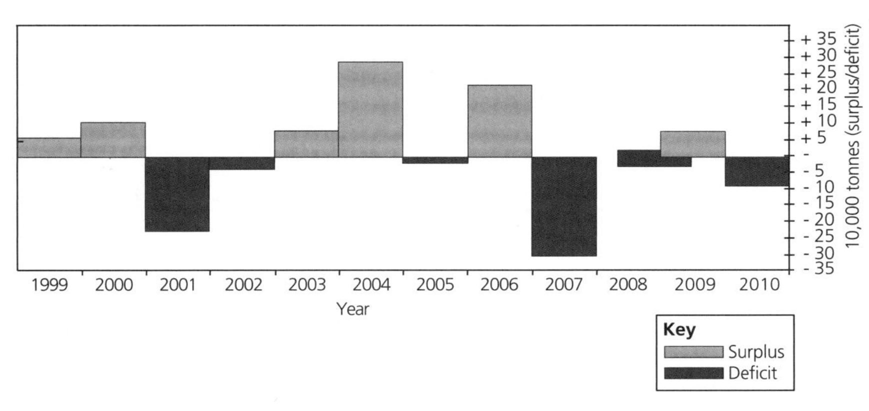

Figure 24 Changes in cocoa supply against demand

Exam practice

a) Describe the pattern of cocoa deficit and surplus shown in Figure 24. Use figures in your answer. [3]

b) Explain how the pattern you have described is likely to affect the farmer's quality of life. [6]

Answers online

Online

Fair trade: a more equal world?

Revised

Fair trade organisations work to improve the lives of farmers, and try to get a better price paid for primary products like cocoa. They don't stop there, though – they also work in the farming communities to help develop sustainable improvements to people's quality of life.

Fairtrade products (certified by the Fairtrade Foundation) are proving increasingly successful. Sales of Fairtrade chocolate in the UK rose from £5 million in 2001 to £343 million in 2010.

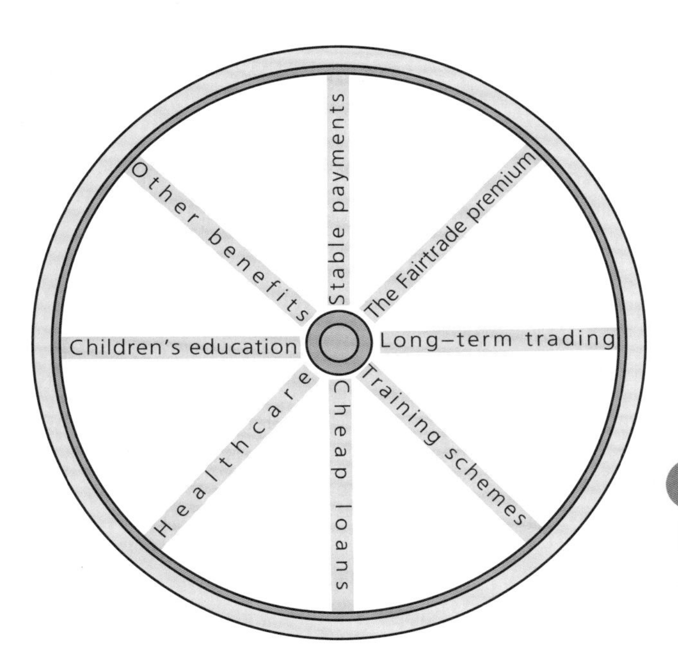

Figure 25 Features of fair trade

Knowing the basics

Look at Figure 25. How might the activities shown in each spoke of the wheel help bring sustainable improvements to the lives of people living in a cocoa farming village in Ghana?

Revised

What is the role of aid?

It is not always possible for countries to respond to events such as natural disasters or war without help. In these situations, help is given in the form of aid. The money and other resources come from three main sources.

Knowing the basics

Read this page and then highlight the article below in four different colours to identify causes and effects of the cholera out break and short and long term solutions.

Revised

Types of aid

Revised

Bilateral aid Aid given by the government of one country to the government of another.

Multilateral aid Aid given by governments to large international organisations which then decide how the aid should be distributed.

Non-government aid Aid given by independent organisations, often charities, which collect donations to use in countries and groups that need help.

Emergency aid Help given to affected areas immediately following a disaster. It concentrates on providing basic necessities like food, shelter and medical help.

How aid is used

Revised

Aid may be used for two broad purposes:

1. As a short-term measure to help people in an emergency.
2. As a long-term measure, in the form of development aid, to help people take control of their lives.

Emergency aid

Oxfam calls for a rapid scale-up of aid effort from governments and individual donations to tackle Sierra Leone's escalating cholera outbreak

Almost 12,000 people have contracted the disease across 10 of the country's 13 districts, with 217 deaths recorded as of 20 August, giving a cholera fatality rate of 1.8 per cent, beyond the World Health Organisation's (WHO) emergency threshold of 1 per cent. The WHO expects that more than 32,000 people are likely to contract cholera during this outbreak.

Grace Ommer, Country Director of Oxfam in Sierra Leone, said 'The cholera outbreak is devastating the lives of vulnerable people such as women and child-headed families, disabled and HIV positive people and the elderly. Cholera is a highly preventable and treatable disease, and yet people are dying across the country due to a lack of safe drinking water, inadequate sanitation, and insufficient access to quick and effective medical care. Unless the humanitarian community steps up its response, cholera will continue to endanger more lives.'

Oxfam is currently reaching 67,000 people in Freetown through emergency water chlorination and is seeking US$4.2 million to help up to 500,000 people with cholera prevention kits, water purification kits and public information campaigns.

Ten years after civil war finished, Sierra Leone is still rebuilding and requires substantial investment in water and sanitation. Only 57 per cent of Sierra Leoneans have access to safe drinking water, and only 40 per cent have access to a private or shared latrine, leaving the majority of the people vulnerable to water-borne diseases. Sewerage systems are also extremely limited.

'The immediate priority is to ensure that people are able to prevent themselves and their families from contracting cholera, or get rapid treatment when they are affected. However, the country needs to tackle this cyclical nature of cholera. Massive investment in sanitation and water is needed to ensure an outbreak of this magnitude never happens again,' said Ommer.

Source: Oxfam News Bulletin, 23 August 2012

↑ **Figure 26 Emergency aid**

Development aid: which route to take to sustainability?

Revised

There is no simple answer to this. In poorer countries, aid can be used to invest in a few high technology development projects or many small scale intermediate technology projects. Supporting farming communities by ensuring a reliable water supply, adequate education and healthcare in each village may be contrasted with importing high yield seeds, artificial fertiliser and hi-tech machinery to grow cash crops for sale abroad.

The two examples below are of aid from the Canadian International Development Agency (CIDA) to countries in Africa. Both were started in the early 1970s.

Location: Mali

Aims:

- to improve village access to basic services
- to provide high quality basic education especially for girls
- to improve healthcare provision in the villages
- to provide more work opportunities
- to enable the provision of fair loans to farmers.

Some effects:

- Faso Jigi: an organisation with 5,000 members in 134 co-operatives. It provides loans and guarantees a fair price and stable income to farmers.
- Sebenikoro Community School: 512 pupils aged 5–12 of whom 320 are girls. All teachers are male.
- National Health Sciences Training Institute: improves the effectiveness of nurses, paramedics, etc.

Location: Tanzania

Aims:

- to set up large scale wheat growing on traditional grazing land
- to export 50 per cent of the wheat and use the rest in a Canadian-operated bakery in Tanzania
- to use machinery imported from Canada
- to provide aid to Tanzania in return for trade with Canada.

Some effects:

- Displacement of the Barabaig pastoral nomad tribe from their traditional lands.
- Provides a source of bread consumed mainly by the urban rich.
- Biosciences Eastern and Central Africa (BECA): helps poor farmers improve their use of farming technology.

Exam practice

As you already know you will be asked in Unit 2B, the problem solving paper, to assess the views of different groups of people and then express your own opinions.

Use this opportunity to practise. Read the information above about the aid provided by CIDA as well as the quotations below.

'... if you ask a Malian farmer what he needs, he will tell you he needs a plough, a pair of oxen and water to irrigate his field. He will not tell you that he needs genetically modified seed.' Ibrahima Coulibaly, Malian farmer.

'Biotechnology will be the key to providing more food and other agricultural commodities from less land and water in the twenty-first century.' Mokombo Swaminathan, Indian geneticist.

1. Which of the two aid programmes above best meets the views of Ibrahima Coulibaly and which the views of Mokombo Swaminathan? Explain your choices.
2. With which view do you agree? Use detailed information to help you explain why.

Answers online

Online

Development issues and water

Drinking water and sanitation

Revised

'Many people take water for granted: they turn on the tap and water flows. Or they go to the supermarket, where they can pick from a dozen brands of bottled water. But for more than a billion people on our planet, clean water is out of reach. And some 2.6 billion people have no access to proper sanitation. The consequences are devastating. Nearly 2 million children die every year of illnesses related to unclean water and poor sanitation – far more than the numbers killed as a result of violent conflict. Meanwhile, all over the world, pollution, over-consumption and poor water management are decreasing the quality and quantity of water.'

Figure 27 Kofi Annan, Secretary General, United Nations 1997–2006

Source: Human Development Report, 2006

'This report contains the welcome announcement that, as of 2010, the [Millennium Development Goals] target [to halve numbers without access to] drinking water has been met. Since 1990, more than 2 billion people have gained access to improved drinking water sources. This achievement is a testament to government leaders, public and private sector utilities, communities and individuals who saw the target not as a dream, but as a vital step towards improving healthcare and well-being.

Of course, much remains to be done. There are still about 780 million people without access to an improved drinking water source. And even though 1.8 billion people have gained access to improved sanitation since 1990, the world remains off-track for the sanitation target.'

Figure 28 Ban Ki-moon, Secretary General, United Nations, 2007–present

Source: Progress on drinking water and sanitation, UNICEF/WHO, 2012

Knowing the basics

1. What does Kofi Annan say about differences in water use in different parts of the world?
2. Ban Ki-moon reports that progress has been only partially successful. What evidence supports this statement?

Revised

Stretch and challenge

Why is water a 'development issue'?

Revised

Contrasts in water supply in the UK

Revised

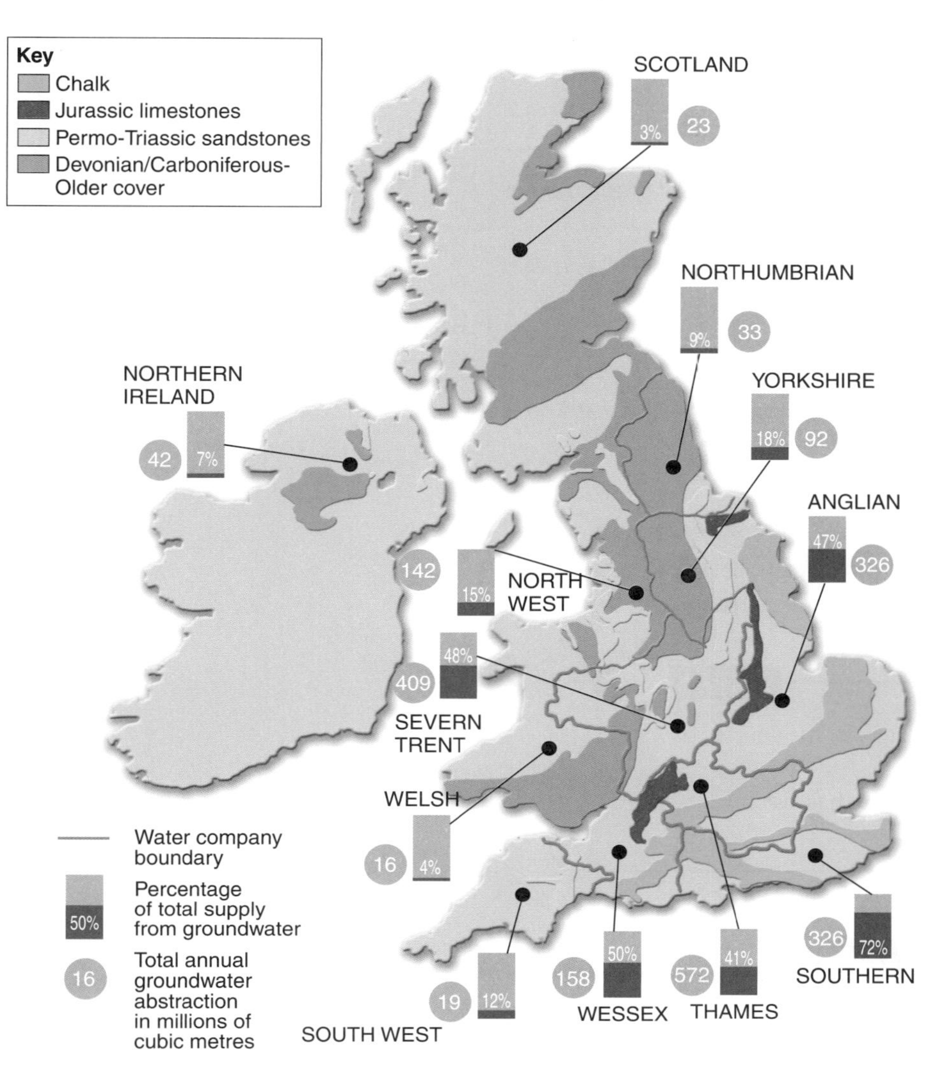

↑ **Figure 29 Major aquifers in the UK. Located bar graphs indicate the percentage of water supply that comes from groundwater supply. The remainder will come from surface stores (rivers and reservoirs)**

Aquifer – a large store of underground water, usually contained in porous rocks

Groundwater – water in the ground below the water table

Knowing the basics

Revised

1. Which water company supplies you?
2. What proportion of water does your water company take from groundwater?
3. In which rocks is this water stored?
4. From where does the remainder of your water supply come?
5. To what extent is your supply of water reliable in terms of quantity and quality?

Contrasts in a poorer country, the Philippines

Source	% of population	Main uses	Comments
a) Street sellers	4	All purposes (drink, cook, wash)	Most live in isolated places with no other choice
b) Public well	35	All purposes	–
c) Private well	15	About half of all purposes	Half for non-drinking water only. Get drinking water from connected neighbour
d) Public standpipe	8	Two-thirds of all purposes	One-third for drinking. Get water from public well for cooking, washing
e) Neighbour connected to piped water supply	38	About half of all purposes	Half for only drinking and cooking. Public well for other purposes

↑ **Figure 30 Cebu, Philippines: water sources of people not connected to the mains water network**

Source: adapted from Human Development Report, 2006

In 2010, safe, piped water reached between 50 and 60 per cent of households in Cebu's urban areas. Its supply was unreliable in terms of both the hours it was available and the pressure under which it was delivered.

Knowing the basics

Revised

1. Suggest a rank order in terms of safety of the sources of water labelled b) to e) in Figure 30.
2. Suggest why only 'isolated' people buy their water from street sellers.

Urban and rural contrasts

Revised

In most of the world's poorer countries, people living in urban areas are more likely to have access to a source of safe, piped water than people living in rural areas. However, that is not to say that urban dwellers don't suffer from water-related problems.

The greatest urban concerns are found in the squatter settlements, which are rarely connected to either mains water or sewerage systems. This is partly because of the nature of such areas. Local authorities find it almost impossible to keep up with their development because settlements are formed so often. They are also illegal and, as such, the authorities are usually uninterested in improving the facilities within them. Consequently, their residents are prone to similar illnesses as those experienced in many rural areas. This is compounded by the very crowded living conditions.

There are many water supply-related killers, including malaria: the anopheles mosquito, which spreads the disease, breeds where there are areas of standing water. Diarrhoea caused by unsafe water is responsible for about 17 per cent of deaths of all 5-year-olds. Cholera is yet another killer disease caused by contaminated water.

Stretch and challenge

Revised

Ban Ki-moon reported mixed progress in achieving the Millennium Development Goals for drinking water and sanitation.

How might 'government leaders, public and private sector utilities, communities and individuals' be persuaded that the same progress is needed for both elements of the target?

Water schemes

Choices: 'small is beautiful' or 'bigger is better'

Revised

Small-scale or large-scale water schemes? The debate continues. As you have seen on page 90, governments, large banks and MNCs still finance expensive multi-purpose water schemes that operate at a very large scale. At the other end of the spectrum are small scale strategies that target individual villages. The British economist 'Fritz' Schumaker brought to the world the phrase 'small is beautiful' as a direct challenge to the 'bigger is better' idea of large schemes.

Case Study: Water, sanitation and hygiene education in Ghana

Revised

WaterAid

Access to water and sanitation facilities in Ghana is low, particularly in rural areas. Only 50 per cent of the rural population in Ghana has access to clean water.

The main sources of water in many parts of rural Ghana are small ponds and unprotected wells, both of which are easily polluted, causing disease and ill-health. Oxfam is working with WaterAid, UK and a local partner, Rural Aid to provide hand dug wells with pumps and constructing ventilated pit latrines.

Nyama Akparibo, aged 28 lives in Asamponbisi Village, Eastern Ghana. WaterAid and Rural Aid helped to build a well in her village.

'I have lived here for 15 years. I have three children, aged ten, eight and six. The water pump was put in three years ago. Our community helped the "water people" (Rural Aid) build the well. The man did the digging and we women collected the sand and stones, helped clear away the sand when the well was dug and helped carry the mortar for construction. We needed a well because we used to drink bad water and get sick, especially the children. We got diarrhoea and stomach pains and felt very weak. We used to be so weak we couldn't do any work. Our main work here is basket weaving, but when you have diarrhoea you are too weak to sit and weave.

'I used to collect water from a stream over there. Animals used to drink form there too and the water was not very clear. When I had my first baby I had to use water from the stream and the baby and I were always sick. Now my children no longer get diarrhoea and we have enough water to bath our children before they go to school, prepare food for them, wash and do our chores. It used to be very time consuming to collect water at the stream as others were already there and you had to queue while they filled their pots before you could collect your own. Now we can just come to the pump and there is always plenty of water.'

http://www.oxfam.org.en/development/ghana/hygiene-education © Oxfam international

The 2011–15 Mekong River Basin Development Strategy

Revised

September 2011

Many rivers flow across international boundaries and sometimes form the boundaries themselves. In these situations the actions of one country concerning the river are likely to have consequences for other countries along its course. Joint agreement and action is very important.

1 Current situation

- not enough water is stored to distribute it between wet and dry seasons
- little groundwater is used in the river basin
- dry season farming is limited by low discharge rates
- only 10 per cent of HEP potential being used
- response to major floods is mainly by soft engineering methods
- water quality is good except in the heavily populated delta
- forests are being destroyed by logging and other land use demands.

2 Development potential

- further HEP production on tributaries
- expansion and intensification of irrigated food production
- some further HEP production on the Mekong itself
- development of fisheries, navigation, drought and flood management, tourism.

3 Development principles

- environment/ecosystem protection
- equality between all countries involved
- maintain water flows on the main river
- remove any harmful effects
- maintain freedom of navigating the river
- respond to emergencies.

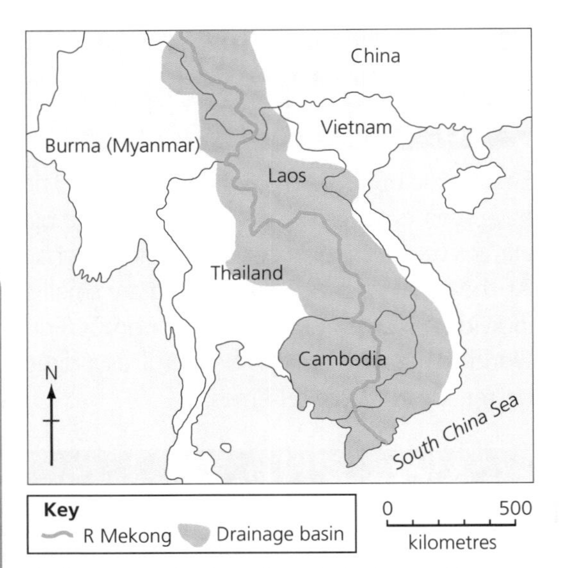

Figure 31 The location of the River Mekong and its river basin

Agreed development priorities: 2011–15

These are the priorities until 2015:

1. To address opportunities and threats of current developments, including closer co-operation with China about its hydro-electric power (HEP) dams. This should improve dry season flow, address sediment rate change issues and provide early flood warning.
2. To expand and intensify irrigated farming, improve food security and increase employment. Introduce improved seed varieties and farm practices. Also put in place drought strategies to ensure greater reliability of water.
3. To improve the sustainability of HEP development by identifying ecosystems needing protection and minimising the negative social effects of damming rivers.
4. To acquire knowledge about current uncertainties and to identify development opportunities. Conduct research into the impacts of changed sediment flows on erosion and deposition, delta formation, habitats and fisheries.

Knowing the basics

Look at Figure 31.

1. Which countries are affected by the River Mekong?
2. Write a 'so what?' statement to help explain each of the facts given as the 'current situation'.

Revised

Stretch and challenge

Read the information about 'development potential' and 'development principles' on Figure 31. Place the 'principles' in rank order according to what you think of their importance. Explain your chosen order.

Revised

Knowing the basics

Read the information about the River Mekong above and on pages 62–65.

1. Why is international management of the River Mekong and its tributaries important?
2. Why is such management often very difficult?

Revised

A different type of examination

This is the final GCSE Geography exam you will take, Unit 2B. You will take it in the same session as Unit 2A, the 30 minute paper on 'Uneven Development and Sustainable Environments'. After you have completed that short paper you will receive Unit 2B.

How will my examination help me?

There are two features of *all* WJEC Geography Specification B exams that are designed to help you fully show your geographical abilities.

1. Resources are used as starting points for each part of the exam. You will either read or complete a resource, before answering other questions designed to help you apply your knowledge and understanding of geography.
2. The exam is stepped. As each new resource is introduced, the level of difficulty is lowered. This is designed to help you work your way through the exam and should boost your confidence. The graph below shows the design of an exam question.

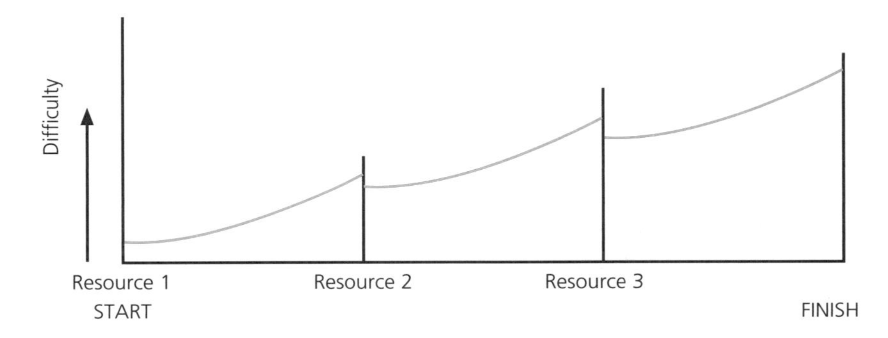

Features unique to problem solving

1. Your examiners really want to know how you would solve a geographical problem. This exam paper will take you through a series of questions, finishing with a chance for you to say what you think and to explain why you think it.

   Take this opportunity; adults don't ask the opinions of young people like you often enough.

2. The problem solving experience is written to test your knowledge and understanding of an issue that cuts across at least two of the three geographical themes. It could ask you to make a straight yes/no decision or, equally, it may ask you to select from or prioritise a number of options.

Issues that have appeared in past papers include:

How should India continue to develop? Improve rural internet access, invest in wind turbines in rural areas, increase university places?

Should a dam be built on a river in Uganda, an east African country?

Should a part of the Yorkshire coast be protected against erosion?

Knowing the basics

Revised

The problem solving paper tests ideas from at least two of the three 'geographical themes' you have studied. Look at the analysis of the question below:

'Should a dam be built on a river in Uganda, an East African country?'

This problem solver looks at the following main areas of geography:

- processes of river erosion (Theme 2)
- effects of people's interference on a river (Theme 2)
- need for water and electricity in a poor country (Themes 1 and 3).

1. Complete a similar analysis for each of the other two boxed questions.
2. Discuss the main areas of geography you have decided upon with a friend.

Tackling a problem solving experience

Over the next few pages you will be taken through a complete problem solving experience. There are two sets of questions, those for the Foundation Tier and those for the Higher Tier. Looking at both sets of questions may help you decide which Tier will be the best for you.

You will probably have experienced many different Foundation and Higher Tier questions in class. Even so, it is a good idea for you to try to answer both papers before deciding the level of exam you will sit. Do you find the Higher Tier just challenging or very off-putting? Is the Foundation Tier experience supportive or just too simple?

The following exam paper looks at the issue of extracting oil from rocks known as shales in the province of Alberta, Canada. It explores different potential effects of such extraction and asks you to decide whether or not the oil should be extracted.

		Marks	
		Foundation	**Higher**
Part A	This introduces you to the issue of oil from oil sands in Canada.	18	16
Part B	This explores, social, environmental and economic effects of the oil extraction.	31	30
Part C	This asks you to decide whether or not oil extraction should continue.	15 [11 + 4]	18 [14 + 4]
	Total marks	**64**	**64**

Foundation Tier – Part A

You are advised to spend about 20 minutes on this part.

This part introduces you to the extraction of oil from oil sands in Canada.

(a) Study Map 1 in the Resource Folder (page 121).

(i) Complete the following passage by circling the correct option in each sentence. [3]

Large areas of oil sands are found in the province of **Alberta / Manitoba / Quebec**.

There are three main areas, the largest of which contains the settlement of **Peace River / Grande Prairie / Fort McMurray**.

From north to south, between points X and Y, this area measures **300 / 400 / 500** kilometres.

(ii) Oil is extracted from oil sands. Oil extraction is in the primary employment sector. Complete the table below to show the sector of other jobs found in the oil industry. [3]

Job	**Employment sector**
Extracting oil	Primary
Hospital working	
Refining oil	
Selling oil	

Exam tips

- You must circle only **one** of the options for each sentence.
- Make sure you attempt **all** the sentences.

(b) Study Graph 1 in the Resource Folder (page 121).

(i) Use information from Graph 1 to help complete the passage below. [3]

The production of oil from sands has ______________________________

between 1980 and 2008 to reach ______________________________

million barrels a day. The rate of change is expected to ______________________________

with output reaching 4–4.5 million barrels a day by 2020.

Exam tip

Make sure that whenever you measure, you do so accurately.

(ii) One way Canada may benefit from the sale of oil is by getting money to spend on public services. Name one public service and explain how it may affect quality of life. [2]

____

____

____

Exam tip

There is 1 mark for naming the public service and 1 mark for explaining how it affects quality of life.

(c) Study the photograph below.

An oil sands facility near Fort McMurray

(i) Use arrows to help you label the photograph to show where the following environmental damage may be caused:

- air pollution
- water pollution
- land degradation. [3]

(ii) Explain two ways in which the damage may affect the quality of life of local people. [4]

Way 1: ______________________________

Explanation: ______________________________

____

Way 2: ______________________________

Explanation: ______________________________

____

Exam tips

- Clearly label the exact places on the photograph.
- You are only asked to label, so don't be tempted to explain.

Exam tip

In all, 4 marks are offered so there are 2 marks available for each 'way', one for a simple statement and one for its 'so what?' elaboration.

[End of Part A: total marks 18]

Part B

You are advised to spend about 30 minutes on this part.

This part examines the social, environmental and economic effects of extracting oil from sands in Canada.

Some social effects of the oil extraction industry

(a) Study the graph below.

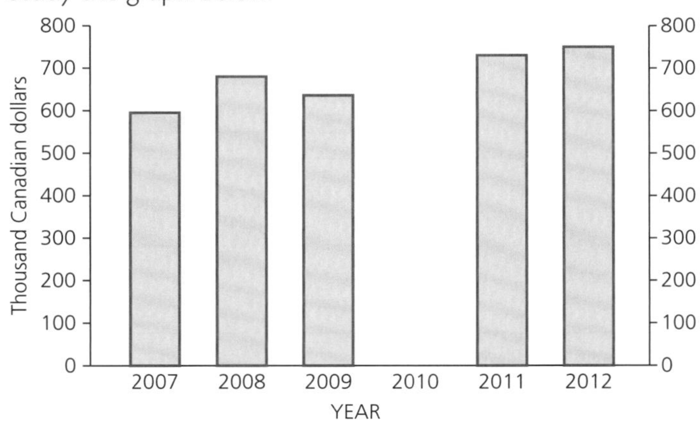

Average house prices in Fort McMurray

(i) Complete the graph on above using the following information. The average house price in Fort McMurray in 2010 was $675,000. [2]

(ii) Suggest how an increase in oil production in the Fort McMurray area may cause changes in house prices. [2]

Exam tips

- Accuracy and the use of a ruler are required here.
- You need to shade the graph the same as the other bars to get the second mark.

____

____

____

(iii) Suggest one group of people for whom the house price changes may be an advantage. Explain why it may be an advantage. [2]

____

____

____

Exam tip

To help answer questions iii) and iv), think back to exercises you have done in class where you consider effects on different groups of people. Consider the groups you discussed then.

(iv) Suggest one group of people for whom the house price changes may be a disadvantage. Explain why it may be a disadvantage. [2]

____

____

____

(b) Study News Article 1 in the Resource Folder (page 122).

(i) One way in which the growth of Fort McMurray has affected the quality of life of local people is through an increase in crime. Explain one effect of crime on quality of life. [2]

Exam tip

You are asked to give one effect and its elaboration, so simple statements about two effects will not do.

____

____

____

(ii) Choose from the news article one other disadvantage and one advantage of the growth of Fort McMurray. Explain each. [4]

Disadvantage: ____

Explanation: ____

____

____

Advantage: ____

Explanation: ____

____

____

Some environmental effects of the oil extraction industry

(c) Study the pie chart below.

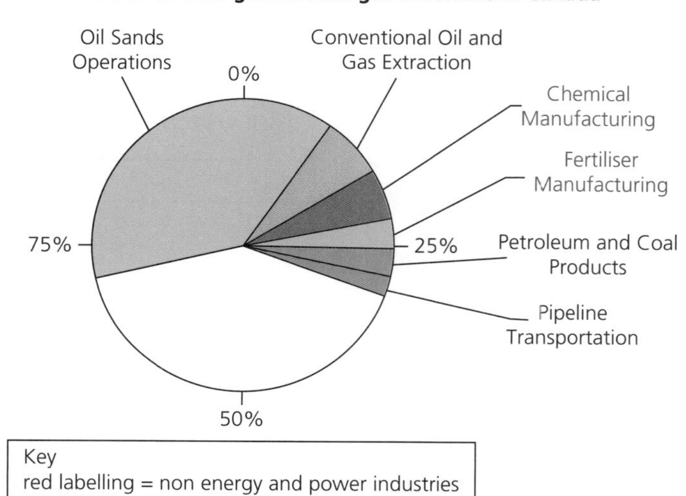

Exam tip

Total the percentages of those emissions typed in red. Keys are not only used on maps in Geography exams!

(i) Complete the pie chart to show the following information about greenhouse gas emissions:

- 37% came from electricity generation
- 4% came from other industries. [2]

(ii) What is the total percentage of emissions from 'non energy and power industries'? [1]

Exam tips

You will need to:

- accurately draw the line separating the two sources *and*
- clearly label them to attract both marks.

____

(iii) Increased greenhouse gas emissions have been blamed for climate change.

Give one feature of climate change.

Explain how each feature may affect the natural environment. [2]

Exam tips

- The breakdown of how the marks are allocated is very clear here!
- Remember, though, that only effects on the *natural* environment will be credited.

Feature: ______________________________

Effect: ______________________________

____

(iv) Give two features of climate change that affect the natural environment.

Explain how each feature may affect the natural environment. [4]

First feature: ______________________________

Effect: ______________________________

____

____

Second feature: ______________________________

____

____

____

Some economic effects of the oil extraction industry

(d) Study Graph 2 in the Resource Folder (page 122).

(i) What is the estimated size of Alberta's oil reserves? [1]

Exam tip

Look back at Graph 2 in the Resource Folder. What does it tell you about the importance of the Alberta oil in relation to other world reserves?

(ii) Suggest why the oil industry may offer long-term employment in the area. [2]

(iii) Explain how the multiplier effect operates in an area of increased job opportunities. [5]

Exam tips

- This is a general question that doesn't need any specific information about Alberta.
- Look back at page 80 for help.
- The question will be marked using a 'levels' mark scheme so the overall quality of your answer and the specific detail you give will be important.

[End of Part B: total marks 31]

Part C

You are advised to spend about 40 minutes on this part.

This part asks you to decide whether or not the extraction of oil from the Alberta oil shales should continue. [11 + 4]

Use the Viewpoints File in the Resource Folder (page 122) to help organise some ideas on the following matrix.

One line of the matrix has been completed for you.

You should spend about 15 minutes completing the matrix.

Exam tips

- Note the marks being awarded. The 11 marks are for the geography of your answer and will be awarded using a 'levels' mark scheme.
- The 4 marks are for the quality of your 'spelling, punctuation and grammar'. This is also 'levels' marked. It is important to ensure that this part of your examination is written in the best English (or for some, Welsh) you can manage.

	Statement	Supports oil extraction (Y / N)
Social effects	Workings destroy large areas of boreal forest	N Because habitats will be lost, threatening some species with extinction.
Environmental effects		
Economic effects		

Exam tips

Completing the matrix

- The matrix is very important to your success in problem solving.
- Its main purpose is to help you organise some ideas, so that you can write a letter or report explaining what you think should happen and why you think this.

A useful backstop

- While you can only get the highest marks for a carefully thought out letter or report, the matrix could be worth valuable marks.
- If you fail to score 8 or more out of the 11 marks available for this part of the paper, your examiner will go back and mark the matrix.

What is it worth?

- A perfectly completed matrix, showing clear geographical understanding in the right-hand column, will be worth 8 marks.

The message

- Make sure you complete the matrix to the best of your ability.
- Include elaborated 'so what?' statements in the right-hand column. Even if you don't end up with a brilliant letter or report, this may still earn you more than two-thirds of the available marks.

When you have completed the matrix, attempt the following task.

Use the information in your matrix to help you write a letter to the Canadian government. You should also use information from other parts of this paper and ideas of your own.

You should advise whether or not the oil extraction should continue. Explain why you have chosen this option in terms of its social, environmental and economic effects.

[End of Part C: total marks 15] You may use up to one more page of lined paper to complete your answer.

Exam tips

To get the highest marks in your letter/report you will need to:

- use elaborated 'so what?' statements
- think about social, environmental and economic effects of the oil extraction and the negative and positive aspects of these
- bring in information from your studies. Perhaps in class you have already looked at some form of mineral extraction and its effects. Use this information to back up your arguments
- consider the short- and long-term effects of the extraction
- be concise. Don't go on and on. If you use any more than the space provided for this task you will likely be just repeating things you have already stated
- write this letter in your best English, or Welsh if your paper is in that language. It will be marked for its spelling and grammar
- use correct geographical terms wherever possible.

Higher Tier – Part A

You are advised to spend about 20 minutes on this part.

This part introduces you to the extraction of oil from oil sands in Canada.

(a) Study Map 1 in the Resource Folder (page 121).

(i) Describe the location of Fort McMurray, a settlement in Canada. [3]

Exam tips

- You are given the information that Fort McMurray is in Canada. Don't repeat it in your answer.
- You could name the province it is in, and the distance and direction from the provincial capital, for your 3 marks.

(ii) In which sector of industry is oil extraction?

Exam tip

Is it primary, secondary, tertiary or quaternary? That's all you need to write.

(b) Study Graph 1 in the Resource Folder (page 121).

(i) Describe changes in oil production between 1980 and those projected for 2020. Use figures in your answer. [3]

Exam tip

You've been asked to give figures and 1 mark will be reserved for their use. If you don't use them accurately, you'll be given a maximum mark of just 2.

(ii) One way Canada may benefit from the sale of oil is by getting money to spend on public services. Name one public service and explain how it may affect a person's quality of life. [2]

Exam tip

There is 1 mark for naming a service provided by local or national government and 1 mark for its effects on an individual.

(c) Study the photograph below.

An oil sands facility near Fort McMurray

(i) Use arrows to help you annotate the photograph to show three different consequences of oil extraction for the natural environment. [3]

(ii) Suggest how damage to the natural environment may affect the quality of life of local people. [4]

Exam tip

When you are asked to 'annotate' you must include an element of explanation in your arrowed statements. For example, don't just label the source of air pollution but go on to describe a consequence of it for the natural environment.

Exam tip

This question will be marked using a 'levels' mark scheme. Give specific detail in your answer to hit the highest level.

____

____

____

____

____

____

____

____

[End of Part A: total marks 16]

Part B

You are advised to spend about 30 minutes on this part.

This part examines the social, environmental and economic effects of extracting oil from sands in Canada.

Some social effects of the oil extraction industry

(a) Study the graph below.

(i) Complete the graph using the following information.

The average house price in Fort McMurray in 2010 was $675,000. [1]

(ii) Explain one reason why an increase in oil production in the Fort McMurray area may cause house prices to increase. [2]

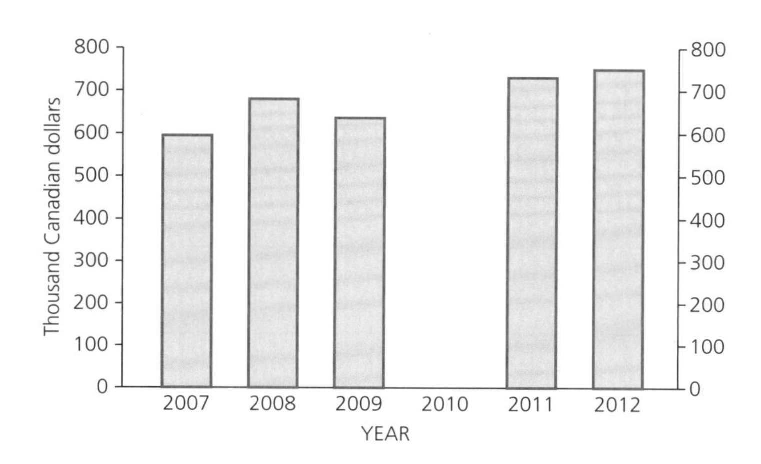

Reason: ______________________________

Explanation: ______________________________

____

____

Exam tip

There is only a 1 mark difference between exam grades. It would be a shame to lose that vital mark by not completing such a simple task!

(iii) Explain how increasing house prices may have different effects on named groups of local people. [4]

Exam tip

There is 1 mark for giving a simple reason and 1 mark for its elaboration.

____

____

____

____

____

____

Exam tip

The key to this question is the term 'named groups'. Use two contrasting groups of people and give a simple effect and its elaboration for each.

(b) Study News Article 1 in the Resource Folder (page 122).

(i) Give one advantage and one disadvantage of Fort McMurray being a 'boom town' based on oil extraction. [2]

Exam tip

You are asked just to 'give' one of each, so don't be tempted to elaborate or explain your advantage or disadvantage.

Advantage: ______________________________

____

Disadvantage: ______________________________

____

(ii) Fort McMurray is described as a 'shift-working' city. Suggest one effect of this on families and one effect on the city as a whole. [4]

Exam tip

Each effect requires a simple statement followed by its elaboration (the 'so what?' statement).

On families: ______________________________

____

____

On the city: ______________________________

____

____

Some environmental effects of the oil extraction industry

(c) Study the pie chart below.

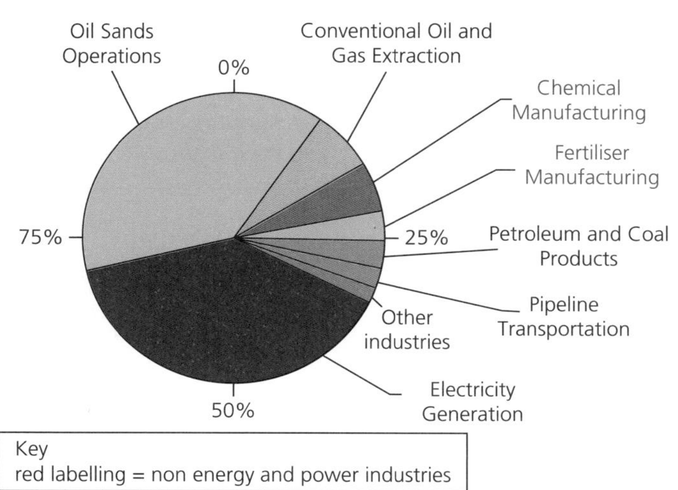

(i) To what extent might the ending of Alberta's oil extraction affect total greenhouse gas emissions? Use evidence from the graph and refer to both the Alberta and Canadian situation. [4]

____

____

____

____

____

____

(ii) Increased greenhouse gas emissions are blamed for climate change.

Explain how climate change may affect natural environments. [4]

> **Exam tip**
> This is quite a complex question. Don't forget to quote figures from the graph and also meaningfully refer to Alberta and Canada as a whole.

____

____

____

Some economic effects of the oil extraction industry

(d) Study Graph 2 in the Resource Folder (page 122).

(i) 'Extracting oil from oil sands will provide sustainable long-term employment opportunities'.

Is this true? Use evidence from the graph to support your answer. [3]

Exam tips

- There isn't a correct answer to this question. You'll receive marks based on how well you justify your choice.
- Merely quoting figures from the graph is not sufficient. You need to translate the evidence into support for the decision you have made.

(ii) Explain how the multiplier effect operates in an area of increased job opportunities. [6]

Exam tips

- This question will be marked using a 'levels' mark scheme. Giving specific detail in your answer will help you to the highest level.
- There are several elements you could use in your response including, for example, the effects of private and public spending on the area. Look back to page 82 for help with this question.

[End of Part B: total marks 30]

Part C

You are advised to spend about 40 minutes on this part.

This part asks you to decide whether or not the extraction of oil from the Alberta oil shales should continue. [14 + 4]

Use the Viewpoints File in the Resource Folder (page 122) to you help organise some ideas on the following matrix.

You should spend no more than 15 minutes completing the matrix.

Exam tips

- Take note of the marks. The 14 marks are for the geography of your answer and will be awarded using a 'levels' mark scheme.
- The 4 marks are for the quality of your 'spelling, punctuation and grammar'. This is also 'levels' marked. It is important to ensure that this part of your examination is written in the best English (or for some, Welsh) you can manage.

	Information from viewpoints file and other sources	**Explanation**
Supports sustainable development		
Does not support sustainable development		

When you have completed the matrix, attempt the following task.

Use the information in your matrix to help you write a letter to the Canadian government. You should also use information from other parts of this paper and ideas of your own.

You should advise whether or not the oil extraction should continue. Explain why you have chosen this option in terms of its social, environmental and economic sustainability.

Don't use any more than 2 pages of lined paper to complete this task.

[End of Part C: total marks 18]

Exam tip

You may feel that you do not need to complete the matrix, which is fine. However, saving the 15 minutes it would take to complete it should not be important in an examination where there are no time pressures. Look back to page 119 to read the advantages of completing it, and make an informed decision whether or not it would be useful to you.

Exam tips

Planning – the key to success

- Carefully consider the options. In this example you are asked to make a yes/no decision. In your actual exam, you may be asked to select from a choice of three options. In that situation, you may be asked to prioritise them.
- Don't just use evidence from the Viewpoints File. Incorporate evidence from other resources and from your wider knowledge and understanding.
- Now work out a structure for your report. It should include:
  - a clear statement of your chosen decision
  - a consideration of the advantages and disadvantages of your rejected strategy with an emphasis on the disadvantages
  - a consideration of the advantages and disadvantages of your chosen strategy with an emphasis on the advantages
  - a final statement briefly restating your choice
  - elaborated responses throughout
  - ideas of sustainability woven throughout.

To reach the higher levels of the mark scheme you will need to:

- use the simple statements in the Viewpoints File and other evidence as starting points. Exploring these will need to trigger more elaborations so you can demonstrate your full understanding
- consider the social, environmental and economic implications of your chosen strategy
- import your own knowledge if it helps to support the views you are expressing
- consider the short- and long-term effects of your strategy. This is a major part of sustainability
- write your report in a formal style. Get straight to the point and don't be tempted to waffle. You should not need more space than the approximate two pages provided for your response
- write in your best possible English, or Welsh if your paper is in that language. It will be marked for its spelling, punctuation and grammar
- include relevant geographical terms wherever possible.

A final word on 'spelling, punctuation and grammar'

- Spelling, punctuation and grammar (SPaG) will be assessed at three places within your overall examination:
  - the case study question for Theme One: 3 marks
  - the case study question for Theme Two: 3 marks
  - the final problem solving report: 4 marks
- An example of a SPaG mark scheme is shown on the right.
- You have been warned! Do not let your inability to express yourself clearly and accurately in the exam undermine your performance.

Level	Descriptor
0	Candidates do not reach the threshold performance outlines in the performance description below.
Threshold performance 1 mark	Candidates spell, punctuate and use the rules of grammar with reasonable accuracy in the context of the demands of the question. Any errors do not hinder meaning in the response. Where required, they use a limited range of specialist terms appropriately.
Intermediate performance 2—3 marks	Candidates spell, punctuate and use the rules of grammar with considerable accuracy and genera; control of meaning in the context of the demands of the question. Where required, they use a good range of specialist terms with facility.
High performance 4 marks	Candidates spell, punctuate and use the rules of grammar with consistent accuracy and effective control of meaning in the context of the demands of the question. Where required, they use a wide range of specialist terms adeptly and with precision.

Resource Folder

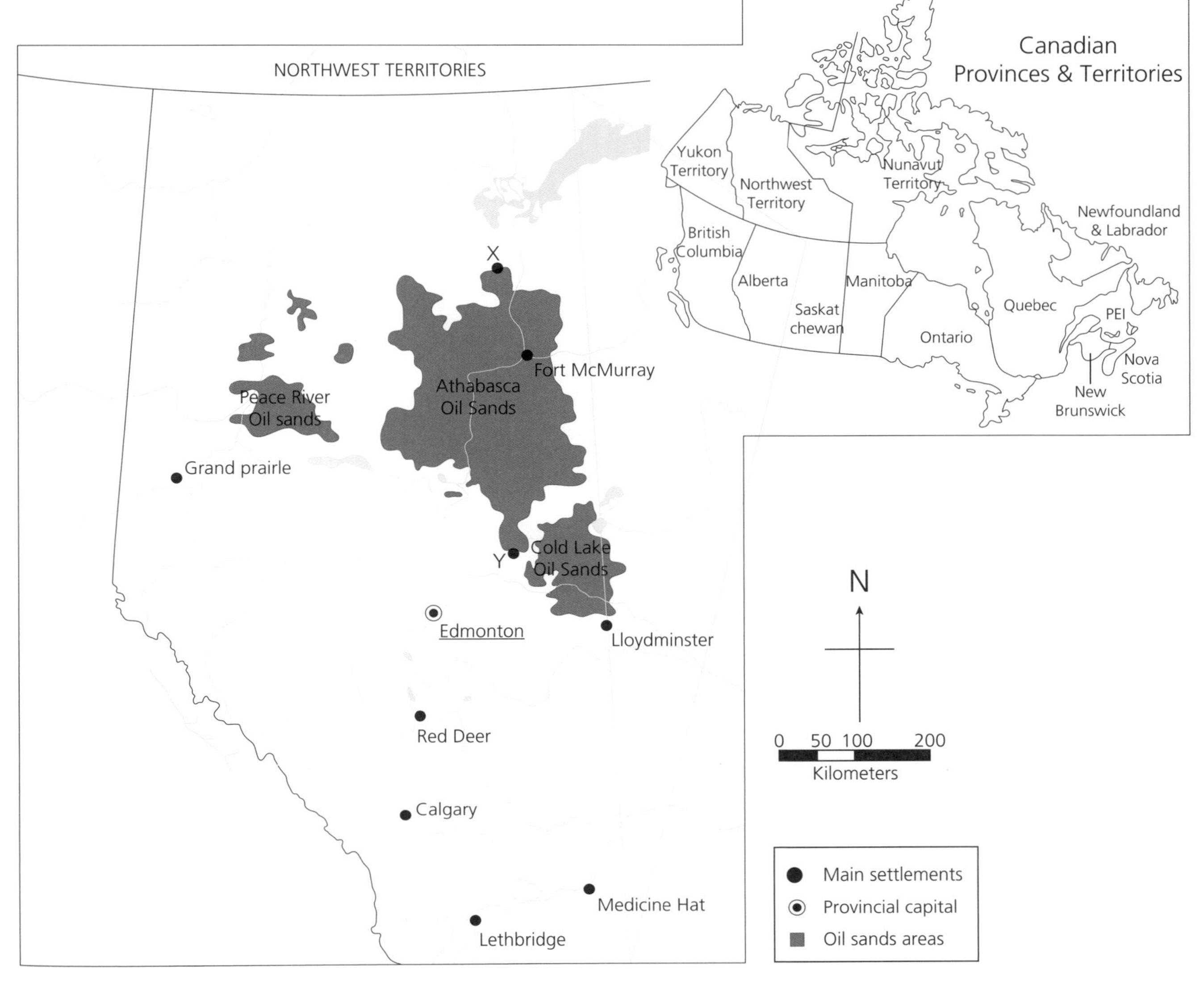

Map 1 Canadian oil sands areas, Alberta

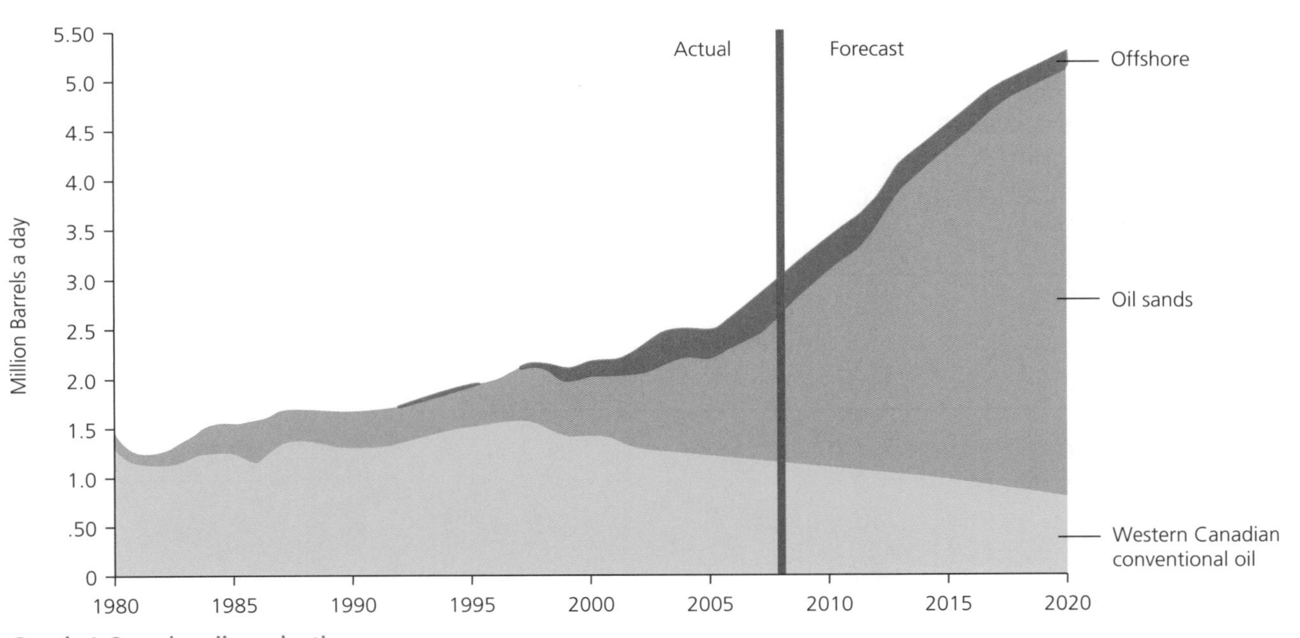

Graph 1 Canada: oil production

Graph 2

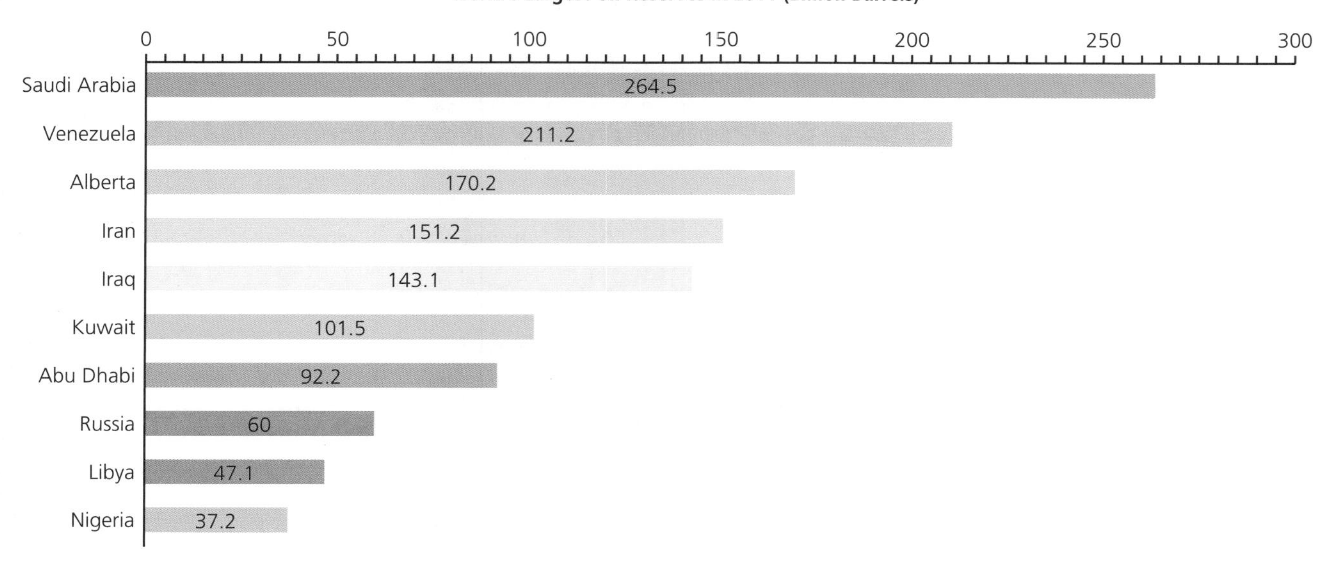

News article 1

The good and bad of boomtown life

Being a boomtown has its ups and downs.

On the upside, Fort McMurray offers excellent cultural experiences. Local theatre, arts, sports, music and other community events take place throughout the year.

On the other hand, there are problems of crime and homelessness. It is also now a 'shift-working city' with oil sands operations taking place 24 hours a day. This brings challenges for working, socialising and family relationships.

Living in a boomtown can be both exciting and challenging.

Viewpoints File

The workings cause too much damage. They disturb large areas of boreal forests, with little done to reclaim them. In cities the sewerage systems struggle to cope with the extra people. A gap is developing in the area between 'rich' oil workers and 'poor' workers in other industries.

It's confusing. Although in the 2011–12 financial year Alberta's government collected about $4.5billion in revenues, the rate paid by the companies is amongst the lowest in the world. Is this enough to make up for the loss of First Nations (American Indian) hunting grounds? On the other hand, they do get jobs in the industry.

We have planted more than 7.7 million trees to replace destroyed forests. More than 90 per cent of all the water used is recycled, preserving water supplies. The Alberta government has invested heavily in healthcare and education in Fort McMurray.

Theme 1 Challenges of Living in a Built Environment

Page 19

Knowing the basics (1)

Remember that quality of life is not merely about possessions and the 'measurable' things in life. So, while negatives feature on the photo may be cramped living conditions, difficulty of reaching the local market or risk of water-borne disease, a positive feature could be the co-operation between the village women in their daily tasks.

Knowing the basics (2)

Indicators on their own are not statements of quality of life. It's the possible effect of these indicators that is important. Thus Ghana, having a much lower GDP than the UK, is likely to have little money to spend on education resulting in a greater proportion of people being illiterate and unable to access better paid jobs or to understand advances in farming techniques. These people would be likely to remain poor and possibly under-nourished. You could follow a similar route through 'healthcare'.

Page 20

1 a) 2e; 3d; 4a; 5c.

b) University student: can't afford a deposit for a mortgage;

recent immigrant: no credit record so won't be offered a mortgage;

newly divorced: probably already repaying mortgage on marital home;

person with frequent career moves: easier to rent if

anticipating moving to another area.

Page 23

Suggested content for the second column of the table is given below:

Missing amenity	Effect on residents
A piped water supply	Danger from water-borne disease
Mains sewerage	Smelly conditions and conditions for pest infestation
Street lighting	Dangers from night crime
Paved roads	Muddy in rainy season and dusty during dry season – uncomfortable and dangerous
A source of electricity	No possibility of using labour-saving devices or entertainment like TV.

Page 24

Suggested answers:

Label	Annotation
Not weatherproof	... so residents could suffer illness from cold and damp
Cramped living conditions	... so no feeling of having personal space or privacy
Difficulty obtaining clean water and difficulty removing sewage	... so wasted time and open to disease.

Page 28

Box 1: Economic and temporary;

Box 2: Refugee and likely to be temporary;

Box 3: Economic and permanent;

Box 4: Refugee and temporary.

Page 31

The push of the countryside

lack of work in area so move because of the perception of employment in urban area;

no school in village so move to get child's education to improve their prospects of a job;

a long walk for medical help so move to urban area with more doctors and nearer hospital;

unreliable water supply so move because they expect a regular piped water supply.

Page 31

1 a) 1e; 2d; 3c; 4b; 5a; 6f.

b) There is no correct answer here. You will need to balance the positive and negative factors and make up your own mind.

Page 32

1 a) and b)

Your list should look something like this:

The young migrant ...

travels alone so has no protection against criminal activity like robbery;

lives outdoors so is open to all weather conditions resulting in illness;

is dependent on charity so has no guarantee of a reliable food supply and may starve;

is moving to an unknown place so is unlikely to have a home or job at the end of the journey.

Page 34

The traditional planning model is mainly a top-down one. Decisions are made by central government and these are acted on at local government level. It is only when decisions are made at these levels that the people who will be affected by the proposed changes will be consulted.

Page 36

Understanding the views of stakeholders in any issue involves trying to put yourself in their position. It is likely, for example, that someone who will have trams passing close to their house will be against the development because of the long-term disturbance of their lives. Others who have had land taken from their gardens or the owners of demolished houses will hold similar negative views. Conversely, the development is likely to be welcomed by people who are not disturbed by it but have the long-term convenience of a tram stop within easy walking distance or by people who live on currently busy main roads that may in future have reduced traffic flowing along them.

One of the main aims of the Nottingham tram network is to reduce the amounts of traffic on the city's roads and, in particular, vehicles entering the city centre. By reducing traffic congestion and air pollution from car exhausts, it is likely to result in a community suffering less mental and physical ill health and a more attractive city where businesses will thrive. In short, it will be a more sustainable community in both the shorter and longer terms.

Page 37

Some of the statements in the table could be either positive or negative depending on how you view them. For example, the Middle Street statement that the route is 'away from the town centre' could be regarded as a positive in that it will keep the pedestrians in the centre safe from the risk of being knocked over by a tram but as a negative in that it will be less convenient for shoppers travelling by tram.

Others are less open to different interpretation. For example, Station Road is often congested. This would result in delayed trams and fewer people using the service due to this unreliability. Similarly, the 'poor bus interchange' of the Middle Street route would not support one of the main aims of the project; that of creating an 'integrated public transport system' for the city. It would also be likely to deter people from using the service.

Question 3 asks you to say which route you would choose. There is no correct answer here and, in a question of this nature, you will gain marks not for the choice you make but for your justification of that choice. Just for the record, the actual route chosen by the developers was Styring Street.

Page 38

Row 2 suggested content: Congested roads > Empty roads > Frustration of hold-ups and not being able to plan when you will reach a destination. Perception of free movement of traffic in rural areas.

Row 3 suggested content: Fear of crime > Friendly community > Feeling that there is great danger of becoming a victim of street crime or burglary in the urban area. Perception of village communities being supportive of each other and having little danger.

Page 39

1 **a), b) & c)**

The village stores will help both economically and socially. It will help economically, for example, by displaying and selling the products of local industries like farmers and craftworkers. This may also attract visitors to the village to buy quality products possibly not available in supermarkets. It will help socially by, for example, enabling pensioners to collect their pensions from the post office and by providing for informal meetings over a coffee and more formal ones in the meeting room. The stores will also mean that people don't need to travel beyond the village for their basic needs.

Page 42

1 & 2 The ticks you place in the table will depend upon how you read the cartoon. However, two of the areas of conflict you may have chosen are:

Motorist and hiker. Reason: Hikers often walk along narrow winding country roads. They are in danger of being knocked down by careless drivers but are themselves a source of frustration for the drivers who often have to wait before overtaking them.

Farmers and picnickers. Reason: Picnickers will often choose a well grazed field as being a suitable place for a meal. They could, however, be in danger of attack from the occupants of the field. By not closing field gates they could allow animals access to dangerous roads and by not clearing away their rubbish, they could put the animals in danger of cuts from sharp rubbish or stomach problems or suffocation from items such as plastic bags.

Page 41

Niagara Falls: the natural feature that attracts people to the location;

hotels: built features that allow people to stay close to the attraction as opposed to having to visit on a day trip. Opens the attraction to visitors from a much larger geographical area.

Cafés: these are a similar convenience to visitors as the hotels;

Dual carriageway: this increases accessibility of the attraction to visitors by car and coach.

Page 42

Free talks on geology = promoting public enjoyment and understanding;

Closure of areas to visitors = conserving and enhancing the natural beauty of the National Park;

A new quarry to be opened = fostering economic and social well-being of communities.

Many of the demands placed on National Parks conflict with each other because they are used by so many different groups of people. Not all decisions made will please all of the users. For example, allowing the opening of a new quarry will ensure employment for local people but will create noise and dust pollution that will upset visitors to the area and owners of second or retirement homes. The National Park Planning Board must also consider the longer term sustainability of the area not just in terms of the environment but also its society and economy.

Page 43

2 Natural features that attract people to Rock include the turquoise water, sandy beaches and good surf.

Built features that attract people include a golf course, cafés and restaurants.

3 Advantages of having a large number of properties as holiday homes include providing trade for local shops and employment in the local leisure based industries like the golf course, cafés and restaurants. However, much of the trade and, therefore, employment is likely to be seasonal.

Disadvantages are mainly effects on the traditional inhabitants of the area. They are being priced out of the housing market.

This is likely to result in younger family members being forced away from the area with resultant split families and an ageing population.

Page 45

Knowing the basics (1)

1 **b)** From left to right: 4; 5; 2; 3; 1

Knowing the basics (2)

Air sinks > it warms > warm air holds more water vapour > no cloud formation > no precipitation.

Page 47

Knowing the basics (1)

1 Effects of a winter anticyclone on a farmer will very much depend upon the activity to be carried out. However, in the case of this farm, conditions may be too cold for grazing animals and they will need to be sheltered. Severe conditions may result in the death of livestock.

On an arable farm it is possible that root vegetables may be frozen in the ground resulting in loss of income for the farmer and shortages in the shops resulting in higher prices.

Knowing the basics (2)

Suggested content for 'Conditions of weather events' table:

Depression:

Positive impacts: often provides much needed water for farmers.

Negative impacts: wet roads creating hazardous driving conditions. Disruption of individual leisure activities.

Summer anticyclone:

Conditions: stable weather. Period of cloudless skies and little wind.

Positive impacts: creates a 'feel good' factor. Increases leisure activity, especially those that are beach based. Encourages cold drink and ice cream sales.

Negative impacts: can create heatwave conditions endangering lives of very old and very young. Little rain could put water supplies at risk. Sometimes results in 'anticyclonic gloom', long periods of still, cloudy conditions.

Winter anticyclones:

Conditions: stable weather. Period of cloudless skies and little wind.

Positive impacts: creates ideal conditions for winter sports like skiing, boosting the economies of areas like the Cairngorms.

Negative impacts: can create extended periods of temperatures below freezing, resulting in disruption of outdoor sporting events and transport. Accompanying freezing fog causes similar disruption.

Page 49

1 & 2 Sustainable solutions are likely to be those that result in more efficient use of existing water. Thus they would include educating people in the use of water and the use of more efficient irrigation systems. Transferring water from the river Segre may also be considered sustainable providing there is capacity surplus to requirements there.

Unsustainable solutions would involve cross-border water transfers because these may be disrupted as a result of political change. Transfer by tanker and the use of desalination plants may be considered unsustainable on the grounds of high cost and wider environmental damage they may cause.

Page 50

Location of the Sahel

Depth from north to south is between 400 and 700 km.

Length from east to west is 5500 km.

Countries affected include Mali, Niger, Chad and Ethiopia.

Rainfall change since 1900

The highest rainfall was 110mm *above average* in 1963.

The lowest rainfall was 125mm *below average* in 1983.

The rainfall trend between 1990 and 1995 was one of rainfall just below average but with a very dry year in 1992 when it fell to 52mm below average.

Page 51

1 Three ways in which rainwater harvesting gets the most use out of available water: by catching runoff in trenches rather than letting it drain away downslope, by using manure to retain water rather than letting it infiltrate beyond the topsoil and by storing and reusing 'grey' water.

2 All elements of the scheme are possible with the use of basic equipment and at a low cost to put in place and maintain. This makes it a sustainable option for a small farm.

Page 55

2 **b)** Heads and tails: 1b, 2c, 3a.

Page 57

1 **c)** Wildlife corridors will allow migration of species between the National Parks. This will encourage the spread of species only currently present in one National Park to the others, thus increasing their biodiversity.

Page 61

A2, B3, C4, D1.

A4, B7, C2, D6, E1, F3, G5.

Page 69

1 From left to right: Niagara Gorge; collapsed limestone blocks; limestone cap rock; less resistant shale.

Areas of undercutting should be labelled underneath the cap rocks of both the American and Horseshoe Falls.

Your annotation should show further wearing back of both the American and Horseshoe Falls and extension of the Niagara Gorge towards the right of the photo.

Page 70

1 Erosion operates on the outside of the meander. Hydraulic action and abrasion undercut the bank which will progressively collapse as a result of gravity.

Material transported by the river will be deposited where the flow of water is slowest on the inside bend of the meander. The more coarse material being transported will be deposited first. This will build out into the river channel.

Page 72

1 A5, B2, C3, D1, E4.

2 Deepening and stabilising a beach is also of commercial value. Holidaymakers are attracted by the beach and occupations catering for these visitors will employ local people. This will, in turn, result in more money being paid in local taxes. This is an example of the positive multiplier effect in action.

Page 73

1, 2 & 3 This is a task that has no correct answer and, in common with the final task of the problem-solving paper, it is asking you to justify your choice of strategy.

Those in favour of holding the line would include local residents whose homes or land would disappear without protection. It is not just the cost of losing a home but many people earn their living from the land that will disappear: hotels, caravan sites, farmland. There is also the sentimental attachment to the family home.

Those in favour of managed retreat may be the local council and other agencies for whom protection would be costly. It is also likely to include residents and landowners of property down drift from the area. Protection here is likely to result in accelerated erosion in these down drift areas placing the property at risk much more rapidly than would have been previously predicted.

Page 74

Knowing the basics (1)

1 Other features of formal employment may include having decent working conditions and set grievance procedures if there is a dispute with the employer.

Other features of informal employment may include only verbal agreements between employer and employee and poor working conditions.

2 Examples of formal employment include working for national governments and for MNCs.

Examples of informal employment include street traders and home-based workers and labourers.

Knowing the basics (2)

1 Primary: farmer, fisher, miner;

Secondary: steelworker, car assembler, silicon chip maker;

Tertiary: nurse, secretary;

Quaternary: IT consultant, medical researcher.

2 In most countries it is likely that such jobs as farmer, fisher, steelworker, miner, car assembler and silicon chip maker will be in the private sector. However, it is possible that any of these jobs could take place in industries that are state owned. The other jobs in the list are even more difficult to place. For example, is the nurse in a private or state hospital or the medical researcher with a pharmaceutical company or a country's national health service?

Page 75

Knowing the basics (1)

No longer retiring at 65 could keep the individual mentally active for longer with positive health implications. It could also result in greater financial security. On the other hand it could deprive younger people of a job.

Lack of education for girls could result in them being tied to the home with no opportunity to experience wider society. This may perpetuate situations in which they are considered inferiors as a result of their gender and are not involved in the family decision-making processes.

With a small proportion of females involved in decision-making at the national level, there is little incentive to improve the status of women as a whole in the country.

The employment of young children deprives them of what we would consider to be a 'normal' childhood and would also ensure that they did not have access to education, meaning that they would be unlikely to break out of the very poor work conditions they have.

Knowing the basics (2)

Tertiary industry increasingly involves the use of IT. Such keyboard skills are associated with traditional female employment in many countries. Improved connectivity also enables many people to now work from home allowing many women to combine wage earning with more family-related tasks.

Increased use of machinery in primary industry has resulted in activities such as farming becoming increasingly capital intensive thus releasing both men and women from the land.

Page 81

Positive aspects of MNCs could include direct and indirect employment and the positive multiplier operating in an area, the bringing in of new technology and skills, the raising of a country's profile and improvement in balance of payments.

The negative aspects may include not only the possibility of closure and resultant negative multiplier but also such factors as exploitation of the workforce, the possibility of few top jobs going to local workers and damage to the local environment.

Page 82

Industrial smoke pollution in the 19th century caused lung disease such as bronchitis. This can reduce both the ability of individuals to work and can also shorten their lives.

Smog created by air pollution disrupts transport and may cause the postponement of sporting events.

Fumes from manufacturing industry have similar effects in the 21st century on the lungs of urban dwellers. It's just the locations that have changed from the old industrial areas of Europe to those of Asia, including China and India.

Page 83

From top to bottom: 3, 6, 1, 4, 2, 5.

Page 84

1 The best way to put together this paragraph would be to combine the statements on the diagram in the order in which they are numbered.

2 The answer to this question is found in the boxes labelled A, B and C. All of these human activities are responsible for increasing the amounts of greenhouse gases in the atmosphere suggesting to many scientists that people are responsible for enhancing the natural greenhouse effect.

Page 86

2 Buy local produce so reducing transport emissions and reducing costs to the family.

Use energy efficient electrical equipment, thereby reducing the need to produce as much electricity and resulting in lower family energy bills.

Buy from renewable energy electrical companies and therefore reduce the burning of fossil fuels to create the energy and makes the family feel it is helping tackle climate change although the electricity may be more expensive.

Reducing heat loss from the house reduces energy use and makes the home feel more comfortable.

Buy durable goods which will need replacing less frequently thus reducing manufacturing gases and family purchase costs.

Packing the fridge tightly reduces electricity consumption and keeps produce at a more even temperature.

Walk or cycle more often to reduce the burning of fossil fuels in cars and buses and save money as a family.

Eat less meat as producing meat is much less energy efficient than eating vegetables direct and costs the family a great deal more.

Use public transport to reduce fuel emissions as a full bus creates much lower emissions than the 30 to 60 cars they would replace. Bus fares are also cheaper for the family than the combined costs of fuel and car parking charges.

Page 88

The continents with the greatest proportion of countries with very low incomes are Africa and Asia.

Although out of date, the Brandt Line still tends to separate almost all of the lowest income countries from the others. However, South America no longer fits well to the south of the line and there are other individual countries in Africa and Asia to which this statement also applies.

Page 89

The choice of indicators of development is very much a personal decision but you will have considered the 'why' of your decision when making it. For example, if you chose access to sanitation you should have thought about the impact of poor sanitation on health and comfort; the spread of water-borne disease like cholera and the discomfort of living in an environment that is likely to be smelly and rodent infested.

There is little doubt that the Brandt Line is out of date. All countries have moved on since 1980. Some have developed much more rapidly than others and this is often shown in a variety of human development indicators. Don't forget, however, that the indicators are often for an entire nation and these hide often vast differences between different areas and groups of people within that country.

Page 90

Migration: 1;

Trade: 5;

Investment: 4;

Culture: 2;

Tourism: 3;

Technology: 6.

Page 91

Knowing the basics (1)

1 50% of Ghana's earnings are of unprocessed cocoa.

2 15.5% of Ghana's earnings are of processed cocoa.

Knowing the basics (2)

Threats which are outside Ghana's control: currency exchange rates, cocoa price changes, drought, changed EU cocoa demands.

The factors above suggest that reliance on a single commodity, whether or not it is cocoa, is not a wise policy for any country. It just takes one of these factors to change for a country to suffer a severe economic problem whereas if they had a large variety of exports, the decline of one would have little effect.

Page 93

Knowing the basics (1)

There would be likely to be two main effects on your life if goods and services were not exchanged between countries. The variety of goods you would have access to would be reduced. For example, you would not be able to consume tropical foods and drinks. The cost of many of the consumer goods you buy would also become more expensive, for example, electrical goods and clothing.

Knowing the basics (2)

Nurses will be attracted to the UK by the higher wages on offer and the prospect of better living conditions. As the UK has strong historical links with the Commonwealth of Nations, migration between those countries and the UK is long established. This is further aided by the fact that most Commonwealth countries have English as either the first or second language making it easy for migrants to the UK to be understood on arrival in the country.

Knowing the basics (3)

You are being asked to weigh up the positives and negatives of this migration. To do so you must look at it from a variety of angles.

UK, the destination country

Positives: migration fills gaps created by not having enough nurses trained here and enables the NHS to continue to operate at current levels so helping the health of the people.

Negatives: it has a minor effect on population pressures like need for housing but also creates some language difficulties in hospitals.

The country of origin

Positives: the possibility of money been sent home to family members.

Negatives: the reduction of nurse numbers in a country with low health care indicators and the removal from the country of people whom it has cost them a lot to train.

These must all be balanced against the personal aspirations of the individual migrants.

Page 94

Quotas imposed on many outside goods will stop manufacturers outside the EU flooding the market with cheap imports enabling UK producers to make a fair profit.

Freedom to travel and work within member countries should allow UK workers to search for employment throughout the EU if no work is available at home.

Import duties imposed on many outside goods should make UK goods cheaper to sell in the EU helping to increase sales as there are no duties on goods produced in the EU.

Subsidies for some goods produced in member countries help industries that find it difficult to survive without help, for example, hill sheep farmers. The subsidy guarantees a price that keeps them in business.

Page 95

Stable payments so able to budget with no worries about the effects of periods of very little money coming in;

The Fairtrade premium so receiving more money than it's possible to get from anywhere else;

Long-term trading so the promise of income over a period of years thus offering security to the farmer;

Training schemes so farmers can benefit from the latest developments in production techniques whether of crops or manufactured products;

Cheap loans so able to buy the latest seeds, fertiliser or machinery without the problem of large repayments to a bank or money lender;

Healthcare so capable of working effectively and being more productive;

Children's education so the next generation will farm more effectively or be able to get a job beyond the farm. Something that could be important in large families.

Page 98

Kofi Annan draws the distinction between the plentiful supplies of water for many people in the world and the lack of access to clean water for many others. He emphasises the health implications of not having a reliable and safe water supply.

Ban Ki-moon states that although more than 2 billion people have gained improved access to safe water since 1990, there are still 780 million who don't have access to it. He also states that progress has not been so good in terms of sanitation.

Page 100

1 It's difficult to be definite about a rank order for this exercise but the order you have chosen should:

recognise the likely greater safety of piped water as opposed to that from a well;

recognise that a private well is likely to be less open to contamination, although much will also depend on the depth of the well;

recognise that the origins of the water sold by street sellers may be unknown.

Recognise that buying from a street seller is likely to be the most expensive option so it will be avoided if there is another nearby supply.

Page 102

Knowing the basics (1)

In the area of the map, China, Myanmar, Laos, Thailand, Cambodia and Vietnam are all affected by the River Mekong.

Not enough water is stored to distribute it between wet and dry seasons so both floods and drought still pose major threats to people living on the banks of the river.

Little groundwater is used in the river basin so there is potential for drawing on these sources for irrigation and drinking purposes.

Dry season farming is limited by low discharge rates so productivity of farms is low resulting in little surplus for the farmers to sell.

Only 10 per cent of HEP potential is being used so opportunities for the economic development of the countries are not being fully exploited.

Response to major floods is mainly by soft engineering methods so still allowing the river to flood. Using hard engineering methods, though more expensive, would allow greater use of the flood plain.

Water quality is good except in the heavily populated delta so for much of its course the risk of water-borne disease is low.

Forests are being destroyed by logging and other land use demands so runoff to rivers is more rapid. This, with increased sedimentation resulting from increased soil erosion, is likely to result in increased flooding.

Knowing the basics (2)

1 The River Mekong flows through a large number of countries and the activities of people along its course will have increasingly greater implications for those countries closer to its mouth. Without the agreement of all of the nations through which the river and its tributaries flow, there could be major environmental, social and economic difficulties for many people in the lower course of the river.

2 Such management is often difficult because developments like a new dam and reservoir that could benefit one or two countries along the course of the river may create problems like reduced irrigation water lower down the course. Given the advantages of such a development, it is often difficult to persuade them not to go ahead with it in spite of the problems it will create for other countries.